THE SOUTHERN SKY

Total solar eclipse, Niugini, 23 November 1984. PHOTO: BOB MILNE

The Southern Sky

A practical guide to astronomy

DAVID REIDY & KEN WALLACE

Allen & Unwin
Sydney Wellington London Boston

First published in 1987

Allen & Unwin Australia Pty Ltd
An Unwin Hyman company
8 Napier Street, North Sydney, NSW 2060 Australia

Allen & Unwin New Zealand Limited
60 Cambridge Terrace, Wellington, New Zealand

Unwin Hyman Limited
37–39 Queen Elizabeth Street, London SE1 2QB England

Allen & Unwin Inc.
8 Winchester Place, Winchester, Mass 01890 USA

National Library of Australia
Cataloguing-in-Publication

Reidy, David.
The southern sky.

Bibliography.
Includes index.
ISBN 0 04 300094 0.

1. Southern sky (Astronomy)—Observers' manuals. 3. Astronomy—Amateurs' manuals. 3. Astronomy—Observers' manuals. I. Wallace, Ken. II. Title.

523

Set in 9/10pt Times by Graphicraft Typesetters Limited, Hong Kong
Printed in Malaysia by SRM Production Services Sdn Bhd

Contents

Figures and tables

FIGURES

TABLES

Introduction

Astronomy is the oldest of the sciences. Since the dawn of time, people have looked at the sky and wondered about what they saw. Even now, when spacecraft have visited most of the planets, and humans have walked on the surface of the Moon, interest in our night skies still abounds.

The southern hemisphere is by far the best place to practise astronomy—we have the centre of the galaxy practically overhead, the brightest stars and the best deep-sky objects. Because of the much lower population it is not hard to find a good, clean, dark observing site from which to practise the hobby. The unfortunate thing about the southern hemisphere is that most of the books written for astronomers are from the northern hemisphere, so the southern objects are likely to be mentioned only in passing or in an appendix.

A more serious hurdle for the beginning astronomer is the fact that many books on the subject either assume too much knowledge or are so simplistic that they are useless. It is to rectify this situation that this book has been written.

When we were planning this book, we decided that there were two main functions we wished to fulfil. The first was to provide a handbook for the beginning amateur astronomer, a book in which all the information necessary to choose and use a small telescope was contained in one place. In all cases we have tried to give practical advice, of use to those starting out. For example, we have given a very simple method of estimating the brightness of variable stars. This method is not used when doing professional observations, but is sufficiently accurate to give you a feel for the work. If you wish to pursue this field further, your local astronomical society or one of the international groups can give you the additional information necessary to take part in their programs.

Our second aim was to provide a comprehensive list of astronomical objects for the owners of small telescopes. Many publications list the non-stellar members of the universe, from the complete New General Catalogue (NGC) to the specialised publications produced by some societies. Many of these otherwise excellent works cause frustration to the beginning astronomer. It is most annoying to read a glowing description of a galaxy, detailing its attractions and features, only to find that you need a 300mm telescope just to glimpse it. The objects we have included in this book are all visible in telescopes with apertures 60mm or bigger, though the larger your telescope, the better the view.

All the objects are also easily visible from the southern hemisphere. We have limited ourselves to objects not much further north than the equator, though there are some small exceptions to the rules. Even with this restriction we think that the selection of objects given is much better than any found in northern hemisphere publications. After all, we do have the best to choose from.

One difference between our book and the publications which come from the United States is the use of the word 'billion'. In this book we have used the English practice and take 'billion' to be 1 000 000 000 000. A book from the U.S. would call this number 'a trillion'. The word 'billion' in the United States and France means 1 000 000 000—in English this is 'a thousand million', or a 'milliard'.

The authors would also like to take the opportunity of thanking all those people without whom writing this book would have been much more difficult. Firstly we would like to thank Joe Cauchi, Glenn Dawes, Steve Burns and Terry Clarke for their wonderful astronomical photographs, and Mike Smith of the Binocular and Telescope Shop, Sydney, for letting us rearrange his store one evening to photograph telescopes and accessories. Proofreading the manuscript was a massive task and we'd like to thank Audrey Reidy and Vincent Joseph for their help in this regard. Finally we would like to thank Steve Jannese and all the other amateur astronomers who gave us support when writing and allowed us to 'pick their brains'.

We hope that this book will help you, the amateur astronomer, to make better use of your telescope and to explore the wonderful universe in which we live.

David Reidy
Ken Wallace
J.D. 2446853

1

Naked-eye observing

Although astronomy has been around for many thousands of years, it is less than four hundred years since 1608 when Hans Lippershey took two spectacle lenses and invented the telescope. It was another year before Galileo Galilei turned this new device towards the skies and started instrumental astronomy. Until this time the only tools the astronomer had were his eyes, yet despite this limitation, many important discoveries had been made; the motion of the planets, the nature of eclipses and even the distance to the Moon were all known to the ancients.

In today's world we have very large, extremely sophisticated telescopes—even some in space, but despite all this amazing technology, the human eye is still the fundamental tool of the astronomer. Whether it be actually looking through a telescope, or just watching the video screen of a computer, an astronomer would be lost without his eyes. Naked-eye astronomy is no longer practised by professional astronomers, but it is a wonderful way for the amateur to get started—there is just so much to see.

Naked-eye astronomy is not simply a case of going outside and looking up; you can do it that way, but it is far more enjoyable and useful if you make a few simple preparations first. To observe well, whether it be naked-eye or otherwise, you must be relaxed and comfortable, and it is to this end that we prepare.

When humans evolved, we developed so that looking ahead or downwards was easiest, looking upwards for any length of time is literally a pain in the neck. The best way to save yourself visits to the chiropractor is to lie down, either on a reclining chair, a mat, or just the ground if it is soft enough, and to look straight up. In this way you will remain relaxed and get a much better view of the skies.

A second problem is the cold. You may think that it is warm and pleasant outside, but when you are observing you will be inactive, and even a light breeze becomes annoying after fifteen minutes or so. It is important to rug up warmly or you will find yourself shivering in no time. For the first few observing sessions it is best to take too many clothes than too few, as it is easier to take a jumper off than to go inside to get one. Once you have observed for a few nights you will know how much clothing is necessary to keep you comfortable and can equip yourself accordingly.

Now that you are reclined and warm, you can start observing. As you do, you will notice something strange happening—the longer you observe, the more stars you can see. This is caused by an effect called 'dark adaptation'. We are all familiar with this effect; if you walk inside after being outside in the bright sunlight, you will known that you can hardly see anything for about a minute—until your eyes adjust. A similar thing happens outside at night; your eyes become more sensitive the longer you wait. This is a combination of two processes.

The first is in the iris of your eye (that's the coloured bit)—as the light level falls it expands the pupil to allow more light in, it then contracts again if you see a bright light. For the pupil to go from small to fully open takes two or three minutes, but dark adaptation takes much longer.

1 This photograph shows the stars in and around the constellation of Orion. Even in this simple shot, it is easy to see the wide range of brightness that stars exhibit. PHOTO: GLENN DAWES

The second process is much slower than the first. It occurs in the part that does the seeing, the retina, at the back of your eye. The retina contains a chemical known as visual purple, and the more of this chemical present, the more sensitive your eye is. In the dark the amount of visual purple increases steadily for fifteen to twenty minutes, giving you complete dark adaptation at the end of this time.

Though it takes up to twenty minutes of hard work to achieve dark adaptation, it takes less than a second to lose it. Your eye is designed to protect itself, so a bright light will cause the visual purple to disappear and the iris to close the pupil. By bright light I mean any light brighter than a star; car headlights, torches, streetlights and even the Moon are bright enough to ruin dark adaptation. When you choose a site, make sure that you avoid all these hazards.

Choosing the site to observe from is probably the hardest thing about naked-eye astronomy; it is difficult to find somewhere that is exactly right, especially in the city. Country sites are by far the best, but most people live in cities or large towns and have to put up with the associated problems. A good city site should be out of range of all lights, either in an area without lights, or in the shadow of something large enough to offer protection, and it should be reasonably protected from the wind. With all this in mind try to find somewhere close, preferably at home, but at least near enough so that carrying books and equipment isn't going to put you off each time you want to go observing.

Now that you're all set to go, what is there to see? The most obvious thing in the night sky is the stars—there are thousands of them. Just how many thousands though? With the dark-adapted human eye and a good country observing site, there are about 6000 stars to be seen in the whole sky. You can't see them all at once though; at any one time you can see only half the sky, or about 3000 stars. In the city it's a different matter—because of 'light pollution' caused by streetlights and other lighting, you can see only 500 to 1000 stars on most nights.

Light pollution is a growing problem around the world—it is the reason that many of the very large telescopes are becoming useless. As towns and cities grow, so does their influence on the environment; cars and industry give out air pollution—buildings and streetlights give out unwanted light. Those lights that illuminate public buildings and monuments at night and look so pretty, make cities the size of Sydney or Auckland visible from many kilometres away. Within the city, becasue of this pollution, the night sky appears dark blue or grey—not black.

Light pollution is the reason why all the new telescopes being built are put in country areas, away from all but the smallest towns. Our skies seem doomed to fade as time goes by. Councils seem to think that there are 'no votes in even simple measures to alleviate the problem.

When you start observing, it may seem that all the stars look the same. Small, colourless pinpoints of light, some of which twinkle, but most of them unspectacular and unchanging. Look again, and the first difference you are likely to notice is the brightness—no two stars are exactly the same. The ancient Babylonian astronomers were the first to try to record the brightness of the stars, and as they had a liking for the numbers six and sixty (it's because of them we have 60 minutes in an hour), they divided the stars into six classes. The top class, class one, contained the brightest stars, class two the next brightest; and so on to class six with the faintest stars. This system, known as the magnitude scale, has remained virtually unchanged for thousands of years.

Over time, astronomers have refined this sixfold system. They decided that a star of a given magnitude had to be 2.51 times brighter than a star one magnitude below it. For example, a fourth magnitude star is 2.51 times the brightness of a fifth magnitude one; and a first magnitude start is 6.30 (that's 2.51 × 2.51) times as bright as a third magnitude one. The problem with becoming so precise is that some stars the Babylonians classified as class one are much brighter than other class one stars—so scientists have extended the range to zero magnitude stars and even stars with negative magnitudes. Sirius, the brightest star in the sky, has a magnitude of −1.6, while the faintest objects ever seen by astronomers (using large telescopes, of course) are twenty-fifth magnitude, over forty thousand million times fainter. Table 1.1 gives you a method of comparing magnitudes and brightness.

The range of brightness visible to the unaided human eye depends upon where you are. A fully dark-adapted eye, in a good, dark, country site, can see to mag. 6.5. Because of the light pollution, only magnitude 4 can be seen from the suburbs, and just magnitude 2 in the heart of a city. The Sun, the brightest object in the sky by far, is magnitude −27 so it is no wonder that you can't see the stars in the daytime.

After you have observed the stars for a while you will notice that they are coloured, not just white. Don't expect bright vibrant colours like a paint chart—the colours of the stars are subdued, not because of the stars themselves, but because of our eyes. The parts of the human eye that sense colour do not work very well unless there is a reasonable amount of light falling on them. Because there is such little light coming from the stars, the receptors are only stimulated a little, and we see only faint colours. In chapter 9 we will discover the significance of these colours, but for naked-eye observing, it is interesting to see just how many different colours you can find. Red and orange stars are very common, as are white ones, yellow stars can be found, and sometimes blue–white. In some places you can find all these clustered very close together, giving a wonderful effect as the colours contrast with each other.

Table 1.1 Method for comparing magnitude and brightness of stars

difference in magnitude	difference in brightness
0	1.00
1	2.51
2	6.31
3	15.85
4	39.81
5	100.00
6	251.19
7	630.96
8	1,584.89
9	3,981.07
10	10,000.00
11	25,118.86
12	63,095.73
13	158,489.32
14	398,107.17
15	1,000,000.00
16	2,511,886.43
17	6,309,573.44
18	15,848,931.92
19	39,810,717.06
20	100,000,000.00
21	251,188,643.15
22	630,957,344.48
23	1,584,893,192.46
24	3,981,071,705.53
25	10,000,000,000.00

Note: To use this table:
1 Subtract the smaller magnitude from the larger one.
2 Look for this value in the left column.
3 Read off the brightness from the right column.

Example:

Magnitude of Sirius = −1.6
Magnitude of κ Tauri = 4.4

4.4 − −1.6 = 6.0

Difference in Mag = 6.0

Difference in Brightness = 251 times.

So Sirius is 251 times brighter than κ Tauri.

Now that you are sitting outside looking at the skies, it is not hard to imagine what it would have been like thousands of years ago, sitting out under the stars discussing the day's hunt, or telling family histories and stories. With no mass media to annoy them, people made up stories about what was around them, the animals, the countryside and, of course, the stars. These early astronomers saw stories in the sky, pictures made up of stars. These pictures were called constellations and they formed a sort of road map of the heavens. Today we still know the constellations,

Taurus, Aquarius and so on, by the names the ancients gave them.

Apart from their mythological significance the stars in a constellation have nothing to do with each other. They simply appear close together in the sky, from Earth. For example, the two stars that form the pointers to the Southern Cross, alpha and beta Centauri, are both in the same constellation, yet they are a long way apart. Alpha Centauri is one of the closest stars to the Earth, while beta Centauri is over a hundred times further away: although they look about the same brightness, beta is actually many thousands of times brighter than alpha.

Another interesting fact is that the constellations are not fixed in shape, they change. Don't except them to change overnight, they won't—in fact they have been the same for all recorded history, but slowly over the ages they will alter. We only know of these changes because of the extremely accurate measurements made by professional astronomers over the past 60 years or so.

Astronomers today still name stars and other objects by the constellation in which they appear, though the system is much more exact than the arbitrary boundaries used by the story tellers. The governing body of astronomy, the International Astronomical Union, or IAU, has divided the sky into 88 areas, each with well defined boundaries—these are the constellations we know today. The constellations of the sky are like the countries of the Earth, we can say that Sirius is the brightest star in Canis Majoris (written α CMa) in the same way as we say Rome is the largest city in Italy. The boundaries, though rigidly defined, are purely arbitrary, and were chosen to correspond to the approximate boundaries of the story tellers' pictures.

The naming of the stars in a constellation is in no way as well defined as the constellation itself. Over the years several methods have been used, resulting in a hotchpotch of systems today. The official method is to give the star its number in the Smithsonian Astrophysical Catalog, but remembering that a certain star is 223 765 is not easy, so most bright stars have names. The systems for naming the brightest stars in a constellation is called the Bayer system. Here the stars are given lower case Greek letters in order of their brightness. The letter alpha is given to the brightest star, beta to the next, and on down to omega for the 24th brightest; then the Bayer system stops. This works well for most constellations, but there were some mix-ups. In Gemini, for example, beta is brighter than alpha, but such exceptions are rare.

Table 1.2 The upper and lower case Greek letters

Α	α	Alpha	Ν	ν	Nu
Β	β	Beta	Ξ	ξ	Xi
Γ	γ	Gamma	Ο	ο	Omicron
Δ	δ	Delta	Π	π	Pi
Ε	ϵ	Epsilon	Ρ	ρ	Rho
Ζ	ζ	Zeta	Σ	σ	Sigma
Η	η	Eta	Τ	τ	Tau
Θ	θ	Theta	Υ	υ	Upsilon
Ι	ι	Iota	Φ	φ	Phi
Κ	κ	Kappa	Χ	χ	Chi
Λ	λ	Lambda	Ψ	ψ	Psi
Μ	μ	Mu	Ω	ω	Omega

After 24 stars a number of methods are used, some involving normal Roman letters, others numbers in catalogues; but these need not concern us here. To complete the name of a star, its constellation name is appended, but not in the form we usually know it. Constellation names are all Latin words so a special form of the name, known as the genitive case, is used. Aquarius becomes Aquarii, Gemini becomes Geminorum, and so on. Table 1.3 gives a complete list of all 88 constellations, along with their genitive cases.

Learning the shapes and positions of constellations is a good pastime for anyone interested in astronomy. Like knowing the streets in the city, it makes it much easier to find your way to places you would like to see. To start to learn your way around the sky find the map in the back of this book that corresponds to the season you are in, then take it outside to observe. You may already have guessed the problem that arises. After all I've said about dark adaptation, how are you going to see the map? There is a way. Don't take your Eveready Dolphin Lantern to use; it's much too bright. Instead, get a small penlight torch and some red cellophane. Cover the end of the torch with three or four layers of the cellophane and secure them with a rubber band. Done properly you will have only a faint red glow barely visible in daylight, but more than enough to read a simple map when your eyes are dark adapted.

When you are outside, start by finding the brightest stars on the map and locating them in the sky. Be patient as it takes some time to begin with. Once you have matched the bright stars, try locating some of the fainter ones. Get used to relating the size of the star on the map to its

Table 1.3 The 88 constellations and their genitive cases

name	genitive	abbreviation
Andromeda	Andromedae	And
Antlia	Antliae	Ant
Apus	Apodis	Aps
Aquarius	Aquarii	Aqr
Aquila	Aquilae	Aql
Ara	Arae	Ara
Aries	Arietis	Ari
Auriga	Aurigae	Aur
Bootes	Bootis	Boo
Caelum	Caeli	Cae
Camelopardalis	Camelopardalis	Cam
Cancer	Cancri	Cnc
Canes Venatici	Canum Venaticorum	CVn
Canis Major	Canis Majoris	CMa
Canis Minor	Canis Minoris	CMi
Capricornus	Capricorni	Cap
Carina	Carinae	Car
Cassiopeia	Cassiopeiae	Cas
Centaurus	Centauri	Cen
Cephus	Cephi	Cep
Cetus	Ceti	Cet
Chamaeleon	Chamaeleontis	Cha
Circinus	Circini	Cir
Columba	Columbae	Col
Coma Berenices	Comae Berenices	Com
Corona Austrinus	Coronae Austrini	CrA
Corona Borealis	Coronae Borealis	CrB
Corvus	Corvi	Crv
Crater	Crateris	Crt
Crux	Crucis	Cru
Cygnus	Cygni	Cyg
Delphinus	Delphini	Del
Dorado	Doradus	Dor
Draco	Draconis	Dra
Equuleus	Equulei	Equ
Eridanus	Eridani	Eri
Fornax	Fornacis	For
Gemini	Geminorum	Gem
Grus	Gruis	Gru
Hercules	Herculis	Her
Horologium	Horologii	Hor
Hydra	Hydrae	Hya
Hydrus	Hydri	Hyi
Indus	Indi	Ind
Lacerta	Lacertae	Lac
Leo	Leonis	Leo
Leo Minor	Leonis Minoris	LMi
Lepus	Leporis	Lep
Libra	Librae	Lib
Lupus	Lupi	Lup
Lynx	Lyncis	Lyn
Lyra	Lyrae	Lyr
Mensa	Mensae	Men
Microscopium	Microscopii	Mic
Monoceros	Monocerotis	Mon
Musca	Muscae	Mus
Norma	Normae	Nor
Octans	Octantis	Oct
Ophiuchus	Ophiuchi	Oph
Orion	Orionis	Ori
Pavo	Pavonis	Pav
Pegasus	Pegasi	Peg
Perseus	Persei	Per
Phoenix	Phoenicis	Phe
Pictor	Pictoris	Pic
Pisces	Piscium	Psc
Piscis Austrinus	Piscis Austrini	PsA
Puppis	Puppis	Pup
Pyxis	Pyxidis	Pyx
Reticulum	Reticuli	Ret
Sagitta	Sagittae	Sge
Sagittarius	Sagittarii	Sgr
Scorpius	Scorpii	Sco
Sculptor	Sculptoris	Scl
Scutum	Scuti	Sct
Serpens	Serpentis	Ser
Sextans	Sextantis	Sex
Taurus	Tauri	Tau
Telescopium	Telescopii	Tel
Triangulum	Trianguli	Tri
Triangulum Australe	Trianguli Australe	TrA
Tucana	Tucanae	Tuc
Ursa Major	Ursae Majoris	UMa
Ursa Minor	Ursae Minoris	UMi
Vela	Velorum	Vel
Virgo	Virginis	Vir
Volans	Volantis	Vol
Vulpecula	Vulpeculae	Vul

brightness in the sky—this will also become easier with practice. Don't try to do the whole sky in one night, take your time and concentrate on just one part, perhaps a quarter. Learn the constellations in this area fully before trying the next bit. In this way you will quickly build up a good working knowledge of the sky's road map.

The interesting thing about the constellations is that they are different every night. Not that they change shape, the shapes are constant over thousands of years, but they are in a different position each night. I think that everybody knows that the Sun rises in the east and sets in the west—well, so do the stars. This rising and setting motion is partly caused by the Earth rotating on its axis once every 23 hours 56 minutes and 4 seconds. A second effect is caused by the revolution of the Earth around the Sun, once every 365¼ days. These two motions combined give the appearance of the Sun, Moon and stars moving across the sky each day.

You will have noticed the difference between the 23h 56m 4s that the Earth takes to rotate and the 24 hours of a normal day. This is because our normal day is measured with respect to the Sun. The 'solar day' is the period from one noon to the next—it combines the Earth's rotation and revolution and is 24 hours long. The stars' apparent motion, on the other hand, is caused only by the

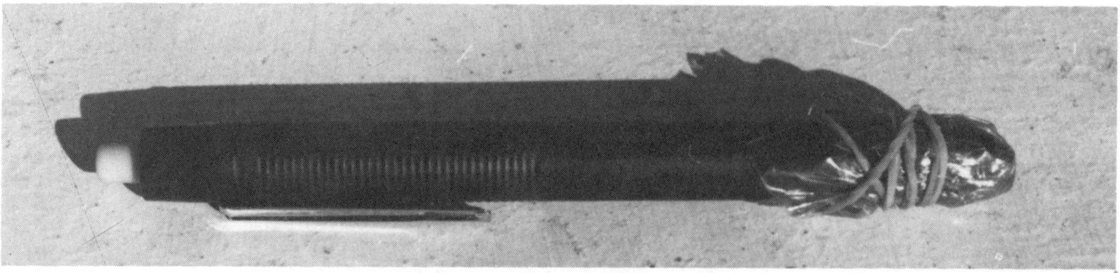

2 A torch, rubber band and some red cellophane are all that is necessary to make a useful device for illuminating maps and other objects at night. By using multiple layers of cellophane, you can adjust the amount of light to suit yourself.
PHOTO: DAVID REIDY

Earth's rotation about its axis, so the period between a start being overhead on two consecutive nights is 23h 56m 4s. This is know as the 'sidereal day'.

A given star rises and sets 4 minutes earlier each day. For example, if Sirius rises at 10:00pm one day, it will rise at 9:56pm the next day, 9:32pm a week later, and 10:00am six months later. Each evening that you go outside to observe you will notice stars in the east a little higher than the previous evening, and stars in the west a little lower. The effect is not large over one night, but over a week or a month it is quite noticeable. That is why we have included times with the maps in this book.

To many people, the night sky is unchanging, some would say boring, without a telescope—yet there is much to see that does change. The planets and the Moon give a different aspect to the sky, night in, night out, but these too are relatively unspectacular without optical aid. Yet every night it is possible to see pieces left over from the formation of the solar system obliterating themselves in a fiery dash across the heavens: the meteors. After only a few nights observing you are sure to see at least one, possible many, meteors streaking across the sky. It always gives me a thrill to see one burning up as it hits our atmosphere at over 100 000 kilometres per hour.

If you could catch a meteor before it destroyed itself, you would find it to be nothing more than a particle of dust about the same size as a sand grain, but let it hit the atmosphere at 30 kilometres per second and it blazes forth. Actually, it is not the meteor itself that glows, but the air around it. Because of its high speed, the sand grain heats the air to many thousands of degrees, causing it to emit light; this is the streak we see. When you have seen a number of meteors, you will notice that no two are quite the same. Some meteors flash briefly across the heavens, others fall slowly leaving long tails behind them, still others break into pieces causing two or more tails; all are spectacular.

Meteors, as I have said, are small dust particles, and it is fair to wonder where all this dust came from. The best guess we have so far is that they are the left over parts of comets, material remaining from when the solar system formed. The main reason we associate meteors with comets is that meteors often come in showers. On any given night you are sure to see one or two meteors in a couple of hours, but during a shower it is not uncommon to see a hundred meteors in an hour. These showers have been shown to follow the same orbit as known comets, the eta Aquariids following the orbit of Halley's Comet, for example. Showers always occur around the same date each year, providing a spectacular show for any who watch.

The best time to view a meteor shower is in the hours leading up to dawn. At this time you are on the front side of the Earth as it moves around the Sun, so more meteors run into the air above you. While Halley's Comet was in our skies in 1986 it was not uncommon to see two or three meteors a minute around dawn, providing a more entertaining view than the comet itself. The best way to see meteors is to lie flat on your back and just look up; any meteors that pass overhead will be immediately visible, and you can follow them along their path quite easily. After a while you will notice that all the meteors tend to come from one area of the sky, a point called the radiant. This is an effect of perspective; the meteors have

3 This meteorite was one of about 500 which crashed to Earth on the evening of 2 September 1967 near the township of Wiluna in Western Australia. The meteorites are believed to have been part of a much larger body, weighing perhaps 250kg which broke up on entry to the atmosphere. The meteorite is made mainly of iron and silicon dioxide. PHOTO: DAVID REIDY AND THE SCOTS COLLEGE PREPARATORY SCHOOL

nothing to do with that region, but they appear to come from there because that is where the Earth's orbit and the shower's orbit cross each other.

Another two phenomena caused by dust that are sometimes visible in a very dark night sky are the 'zodiacal light' and the gegenschein. Zodiacal light is sunlight that is scattered from more of the particles of dust that have been left over from the formation of the solar system. There is not much of this dust left, but what there is appears as a very faint band of light stretching upwards from the eastern or western horizon. Sometimes this glow extends almost halfway up the sky. The gegenschein is a similar phenomenon, it appears as a faintly glowing patch directly opposite the Sun; it too is caused by dust scattered throughout the solar system. This time though, the dust is thought to be caused by the Earth, muck escaping from our atmosphere. The best time to view gegenschein from the southern hemisphere is around midnight during June and July; then it is almost directly overhead.

An object that is associated with meteors is the meteorite. This is a lump of rock or metal that survives its journey through the atmosphere and lands on the Earth intact. If this happens on land, a crater is formed, the size of the hole depending on the size of the meteorite. The very famous Barringer Crater in Arizona, USA formed about 20 000 years ago. It was caused by an object weighing over a tonne hitting the plain at over 30 kilometres per second. There are other meteor craters all around the world, including many in Australia.

Though they have similar names, meteors and meteorites have different origins. Meteors come from comets, but meteorites probably come from the asteroid belt, a collection of rocks and dust that orbits the Sun between Mars and Jupiter. The theory has it that sometimes these particles fall in towards the Sun, occasionally colliding with the Earth. Although there are many potential meteorites in the solar system, the number that hit the Earth is incredibly small, and so far there has never been a case recorded of anyone being hit by one. When it falls, a meteorite is often accompanied by a loud bang—not as it hits the Earth, but caused by the sonic boom as it passes through the air. If you are lucky enough to see a meteorite fall, contact your local museum or astronomical society as they would be very anxious to get it for study.

Another phenomenon in the night sky that is becoming increasingly common is the satellite. The Earth has one natural satellite, the Moon, but more are constantly being put up to facilitate

Table 1.4 Major-meteor showers visible in the southern hemisphere

name	best dates	notes
Corona Australids	Mar 14–18	
η Aquariids	May 1–8	Very fast moving objects
Ophiuchids	Jun 17–26	
Capricornids	Jul 10–Aug 15	Very slow, yellow
δ Aquariids	Jul 15–Aug 15	Slow, long paths. Two radiants
Piscis Australids	Jul 15–Aug 20	
α Capricornids	Jul 15–Aug 25	Yellow
ι Aquariids	Jul 15–Aug 25	Two radiants
Orionids	Oct 16–27	Very fast
Taurids	Oct 10–Dec 3	Slow, brilliant
Phoenicids	Dec 4	

Note: It is impossible to say whether any shower will be spectacular at any one time as they change a great deal from year to year. The best time to look for the meteors is around the middle of the period listed. Some showers put on a good display over a few days, others only for a few hours. The only way to find out about the shower is to go out in the pre-dawn hours and have a look at it.

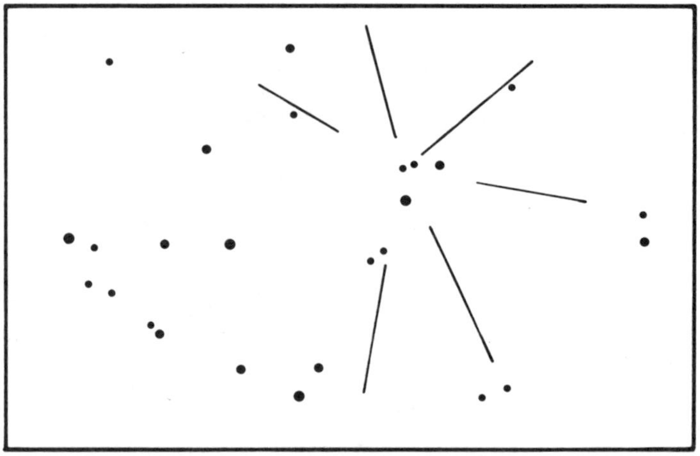

4 The tracks of six meteors of the α Capricornids shower. When plotted in this manner, all the tracks appear to come from the region around α Capricorni though in reality there is no connection. The effect arises from perspective, the meteors really hit the Earth with almost parallel paths.

communications, surveying and spying. Being basically a litter-bug, mankind also has a fair bit of junk in space—bits and pieces ranging in size from burned-out rocket casings to dropped spanners orbit around the Earth. All these objects reflect sunlight and many are easily visible from Earth, moving across the sky at a sedate pace, usually fading from view as they pass into the shadow of the planet. It you want to observe them, the best time is just after sunset and just before sunrise. On any day there are ten or twenty easily visible objects to observe, and as more stuff goes up, so more will be seen.

Occasionally one of these satellites, or pieces of junk will re-enter the atmosphere, burning up in much the same way as a meteor does. Bigger pieces of junk sometimes make it to the Earth's surface intact, as parts of Skylab did in 1980. These events are usually well publicised in advance by the media, and make an interesting sight when they occur.

Naked-eye astronomy will give you many nights of pleasure exploring the heavens in the same way as the first astronomers did. There is always something new to see, something you hadn't noticed before, and as you continue more and more will be revealed. It will also give you a good knowledge of where everything is; this is invaluable for any further work in astronomy.

2

Astronomy with binoculars

Before long you are going to want to see all the objects in the night sky in greater detail, to see them close up. Of course, you can go out and buy a telescope, but there is an easier way. One that is a lot less expensive. Around the home you may well have a pair of binoculars, the type used at the races, or for watching boats. These are an ideal way to get started in instrumental astronomy, as binoculars are much easier to use than any telescope.

It may be that you don't have a pair of suitable binoculars but would like to buy some—after all, even if you give up astronomy they will still be useful for other purposes. When choosing a pair there are a number of simple points to consider. As there are many different styles of binoculars, it is best to go to a shop with a wide range of brands. The wide range will give you more choice and you are more likely to find a pair that is just right.

The most important information about a pair of binoculars is their size, this is in the form 7 × 50, or 8 × 30: two numbers separated by a multiplication sign. This is pronounced 'seven by fifty' or 'eight by thirty' and is usually engraved somewhere on the body. The code is easy to decipher; the first number is the magnification of the binoculars, that is, how many times bigger or closer they make objects appear. For example, binoculars with a magnification of eight, would make a boat appear eight times bigger—or eight times closer—than it is without them. The second number is the size of the front lenses in millimetres; this affects how bright and colourful the view appears. So, a pair of binoculars labelled 7 × 50 have a magnification of seven and a front lens 50mm in diameter.

It may seem sensible to choose the pair of binoculars with the largest magnification—after all, the stars and things are tiny so we want to make them as large as possible—but large magnifications lead to large problems. The most important problem, and one that is not easy to overcome, is the lack of light. The more you try to magnify an image, the more you have to spread the light out. A suitable analogy is buttering bread. If you were to take the amount of butter you normally put on one slice of bread, and try to spread it over 64 slices, you would have the same effect that eight times magnification gives (64 is eight times eight). The butter would end up very thin. When this happens with the light, you end up with a very dim image. The only way to overcome this problem is to start with more butter, or more light. It is here that the aperture of the binoculars becomes important.

When your eye is fully dark adapted, the pupil is about seven millimetres in diameter so it allows a certain amount of light into your eye. Binocular lenses are generally much larger than this, and so they allow a greater amount in. Table 2.1 shows just how much more light. It is this extra light that allows you to use magnification without the image fading so much that you can't see it. It is important to choose the best magnification for the aperture; this is also shown in Table 2.1. If you choose a magnification bigger than the optimum, the image will be fainter than it appears with your eye, if you choose a magnification that is smaller, the image will be brighter. A brighter

5 Three pairs of binoculars commonly used for astronomical observing are seen here. They range from the 10 × 80 pair at the rear to a pair of 7 × 50 and a compact pair of 7 × 30 in front. PHOTO: DAVID REIDY

image is no problem, but a faint one is unacceptable for astronomy.

There is another problem caused by high magnifications. The magnification that any optical instrument gives also magnifies any movement in it. Steadying a heavy pair of binoculars with a large magnification can be very difficult. Most people can manage about seven or eight times magnification without problems, but you can best judge this for yourself by trying several pairs when you are ready to buy. Very large magnification binoculars are normally mounted on a tripod or stand to hold them steady, and most smaller pairs have provision for this too.

Apart from aperture and magnification, several other features should be considered when choosing a pair of binoculars—most of these affect the quality of the instrument and its price. When light strikes any surface some of it is reflected back, as a mirror does; this reflection means you can lose as much as 5% of the light every time it enters a lens. To alleviate this problem, some binoculars have coated lenses. This is a process where a very thin film of chemical, usually magnesium fluoride, is put on the surfaces of all the lenses. This coating reduces the amount of light that reflects from the lens to about 0.1%. Binoculars with this type of treatment will give a brighter image than those without such coating. Some binoculars are even multi-coated to reduce reflections even further and to increase the toughness of the lens surface. Coating adds to the price of the binoculars, but is worth it for the brighter, clearer images it gives.

A second thing to consider when choosing binoculars, is their ease of use. Check that they are comfortable to hold, that you can easily use the focusing knob and that they are not so heavy they will make your arms ache. All these things are separate from the function of the binoculars, but they are important as they affect the enjoyment you will receive from using them. Lastly, beware of very cheap binoculars; they are worthless. A good pair will cost over $50, and pairs costing less than this are usually of such poor quality as to be useless. Some cheap pairs are labelled Galilean binoculars—stay clear of these. Galilean binoculars are the type used in opera glasses and because of their simple optical arrangement they are totally unsuitable for astronomical observations.

A good pair of binoculars will give many years of enjoyment if you take good care of them. When you're outside observing, use the neck strap as it is very easy to drop the binoculars when you are stumbling around in the dark. Because of the temperature drop at night, dew tends to form on exposed surfaces, especially the

Table 2.1 Magnification with binoculars

aperture in mm	amount of light	maximum magnification
20	8	3
25	13	4
30	18	4
35	25	5
40	33	6
45	41	6
50	51	7
55	62	8
60	73	9
65	86	9
70	100	10
75	115	11
80	131	11
85	147	12
90	165	13
95	184	14
100	204	14

Note: To use this table, find the aperture of the binoculars in the left column, then look at the number in the middle column—this tells you how many times more light the binoculars receive than a dark adapted human eye. The right hand column tells you the maximum useful magnification for the binoculars.

metal and glass surfaces of binoculars. It is a better idea to keep your binoculars warm, as this will prevent dew from forming, than to be continually trying to wipe it off. If you find the binoculars dewing up, put them under your coat or jumper for a few minutes and this should dry them out. If they become very heavily dewed, take them inside. Wiping the lenses when they are wet will probably affect their coatings.

The best way to keep the lenses clean is to gently blow dust and dirt off them. You should never rub a lens with anything as this risks scratching the lens or its anti-reflective coating. Of course, never touch any lens surface with your finger; it will leave a greasy mark which is almost impossible to remove. Finally, always store your binoculars in their case with the lens caps on. This will keep dust from the surfaces and will protect the binoculars from any knocks and bumps. In the case you should have a small packet of silica gel; this absorbs moisture from the air to help keep the binoculars dry. Always allow a wet pair of binoculars to dry before storing them. If your binoculars should ever need fixing, take them back to where you bought them or to a qualified technician. Binoculars require very accurate adjustment to work properly, and there is no way to do this without the proper equipment.

Learning to use your binoculars correctly is just as important as buying the correct pair. A quality instrument is only as good as the way you use it. There are a few simple techniques which apply to all binoculars and which, if learned properly, make observing much simpler and more enjoyable. The first of these is focusing. Without correctly focused binoculars, not only will you get a poor view of the heavens, but you will probably end up with eye strain and a headache.

On one of the eyepieces of binoculars is a turning ring called the dioptre adjustment. This is normally found on the right-hand eyepiece, but it is sometimes on the left. People who wear glasses are probably aware that their eyes are different strengths; this also applies, though not as strongly, to people who don't need glasses. The dioptre adjustment lets you set the binoculars so they are exactly right for both your eyes. To use it properly is as easy as following a few simple steps.

6 The dioptre adjustment on a pair of 8 × 50 binoculars. PHOTO: DAVID REIDY

1 Using one of the front lens caps, cover the side of the binoculars with the dioptre adjustment, usually the right hand side.
2 Point the binoculars towards a feature in the far distance, a boat at sea, a distant telegraph pole or the like. At night it is easiest to use a bright star.
3 With both eyes open and relaxed, focus the binoculars, using the main focusing knob, so that you can see the object clearly. If you are using a bright star, focus so that the image is as small as possible.
4 Without adjusting the focus, move the lens cover from the side with the dioptre to the side without it.
5 Again point the binoculars towards the distant object.
6 Using *only* the dioptre adjustment and *not* the main focusing control, with both eyes open and relaxed, focus the binoculars so the view is clear and distinct.
7 Remove the lens cap completely so both lenses are clear and view the object. The view should be perfectly focused for both eyes.

Using binoculars during the daytime is as simple as pointing them at an object, focusing and watching, but for astronomy it is not that simple. Care is needed in choosing a site for binocular observing. Bright lights and wind are

7 This simple bracket screws into the front of a pair of binoculars to allow them to be attached to a camera tripod for hands-off viewing. Having such a device for your binoculars is a great advantage when it comes to long observing sessions. PHOTO: DAVID REIDY

just as much of a problem with binoculars as they are without.

One factor you may have to consider a little more closely is comfort. Looking up for any length of time is difficult enough, but with a heavy pair of binoculars it becomes tiresome very quickly. If you don't already have a reclining chair of some sort, now is the time to get one, preferably one with arm rests. The easiest way to use binoculars is to lie back with your arms resting on the arm rests to help support the binoculars. If you have invested in a very large pair of binoculars, you may find it easier to mount them on a camera tripod or some other stand and to sit or stand separately from them. A stand is certainly an advantage if you want to show other people the sights. You can set the binoculars so they are pointing in the right direction, then all your guests will need to do is to focus them.

One of the first things to try when you are outside with your binoculars, is to look at the stars. You may be surprised. Using a good pair of binoculars, the stars look the same as they do to the naked eye. A quick calculation will show that this is hardly surprising. The closest star to the Earth after the Sun, α Centauri, is 40 billion kilometres away. If you bring that eight times closer it is still five billion kilometres distant, so it is hardly surprising that it still appears as only a point of light. The only difference a pair of binoculars will make to stellar observing is the darkness of the sky. Although the stars will appear about the same brightness, the sky will appear darker. This is a definite advantage for city viewers as it means that more stars are visible due to the increased contrast between them and the sky.

For astronomy, binoculars work best on extended objects, objects like the Moon, the larger planets and comets. As the Moon is always around somewhere, it is a good object to start on. In chapter 8 I will deal in greater detail with observing the Moon with a telescope; for now we will look only at the fundamentals. One important thing to remember is that because the Moon is so bright it will ruin your dark adaptation, so leave it until last each night.

The first thing you will notice when you look at the Moon, is the amazing amount of detail that is visible. With the naked eye the Moon has a blotchy appearance—with binoculars these blotches turn into vast plains lined by mountains; craters hundreds of kilometres in diameter spring into view. Literally, a whole new world is seen. It is easy to be overwhelmed by all the detail, but if you concentrate on a little at a time, eventually the surface of the Moon will be as familiar as a map of our own world.

The best time to begin observing the Moon is not when it is full, and dominating the night sky. At this time the sunlight is falling onto the surface in such a way that there are no shadows visible from the Earth. Because the Moon is pretty much the same colour all over, it is difficult to pick out detail when there are no shadows. A better time is when the Moon is at first or last quarter, then half of the face visible from Earth is illuminated and the shadows are quite noticeable.

If you look closely at the face of the Moon, you will see that a chain of large craters runs down the centre of the southern hemisphere. At the quarter moons it is possible to see the floors of these craters in shadow, while the rim of mountains around them picks up the sunlight—it's a wonderful view. The craters which pock-mark the Moon's face were caused by meteorites crashing into it during the period when the solar system was forming. Because of the lack of atmosphere on the Moon, these marks have not been erased the way they have been on Earth. When we look at the Moon, we see a picture of what the Earth may have looked like 4000 million years ago.

An interesting fact about the Moon which will quickly become apparent is that it always has the same face pointing towards the Earth. This is called captured rotation—it is caused by the gravitational force from the Earth gradually stopping the Moon from rotating. Another thing you will notice about the Moon is that it rises later each day. The Moon takes 28 days to revolve once around the Earth, and this combined with the Earth's rotation makes it appear to rise 50 minutes later each day. Also the phase the Moon shows corresponds to the time it rises. When new moon occurs, the Moon rises at the same time as the Sun; at first quarter it rises around midday; a full moon always rises as the Sun is setting; and last quarter rises at midnight. So to view the Moon through all its phases, you will need to rise at all hours of the night.

When the Moon is a thin crescent, and visible in a dark sky, you may see what appears to be a ghostly image of the missing part of the Moon. This is caused by a phenomenon known as 'earthshine'. Just as the Moon reflects sunlight to the Earth, so the Earth reflects sunlight to the Moon. From the Moon, the Earth appears about four times as large as the full moon does from Earth, so it illuminates the Moon's night time very well.

The Moon is an interesting object, but most of the work you can do with it needs a telescope. The field of astronomy in which binoculars are supreme is cometary observation. Because of their large size and relatively high brightness, comets are ideal subjects for binocular observing. Unfortunately, large bright comets are not very common, but smaller comets are regular visitors to our region of the solar system.

Halley's comet is by far the most famous comet, especially since its visit in early 1986. Many people were disappointed by what they saw when Halley's was in our skies, but Halley's return this time was very good as far as comets go. Throughout the years stories have surfaced about comets being so bright they are visible during the daytime—this does happen, but such comets are quite rare, coming around about once every hundred years. Most comets are too faint to be seen with the naked eye or small binoculars, but every five to ten years a comet comes along that is about as visible as Halley's was.

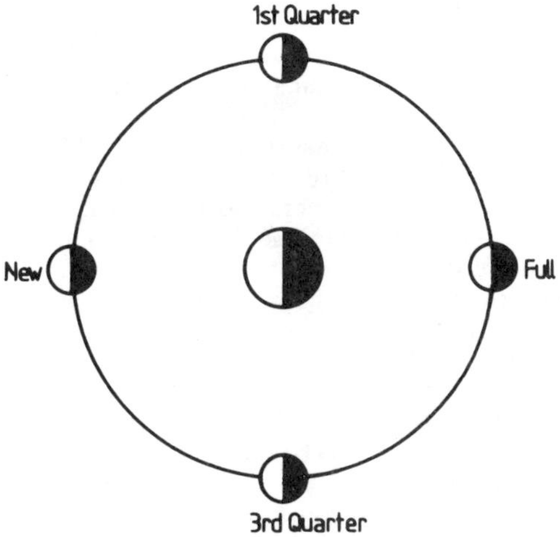

8 A simple diagram showing how the phases of the Moon are formed. In this picture the sun is towards the left, illuminating half of the Earth and half of the Moon.

The techniques used to observe a comet are fairly simple, and you can consult your local astronomical society for information about a new comet. A comet usually appears in the sky in the hours just after sunset or just before sunrise, and the best time to start looking for it is about an hour before astronomical twilight starts or an hour before it finishes. Astronomical twilight is the time after sunset or before sunrise when professional astronomers consider the sky is too light to do any serious work. It is defined to be the period when the Sun is less than eighteen degrees below the horizon, but a good rule of thumb is an hour and a half before sunrise or after sunset. Therefore, you should start to look about half an hour after sunset, or two and a half hours before sunrise.

When searching for the comet it is best to start with the naked eye; in this way you can see more of the sky. Start by letting your eyes adapt to the dark, then begin looking for the comet. Scan across the area slowly—take your time as it often takes a while to find a faint comet. You may find that you catch a glimpse of the comet from the corner of your eye, but when you look to find it, it is gone. You're not seeing ghosts—it is the comet, but it is too faint to be seen using direct vision.

The human eye has two areas of different sensitivity. The part of your retina you use when reading, and for general seeing is called the fovea. It is sensitive to fine detail and colour, but it does not respond well to very low light levels. The other part of your retina takes care of peripheral vision, the edges around your centre of attention. This part sees less detail and colour, but it is much more sensitive to low light levels. To take advantage of this added sensitivity astronomers use a technique known as 'averted vision'. This involves looking slightly to one side of an object while concentrating your attention on the object itself. It is a difficult technique to master, but gradually you will become used to doing it.

Once you have found the comet, either using direct or averted vision, don't take your eyes off it, just bring your binoculars up and look. You should be able to see the comet without difficulty. Before examining what you can see when viewing a comet, it is necessary to understand what a comet is. The best description we have is that a comet is a big dirty snowball. Up until the beginning of 1986 the best information we had about comets came from ground-based observations made with large telescopes and specialised instruments. Since then, six spacecraft have visited comets, so we now know a lot more. Comets are made from ice, dust and organic materials; they are lumps about ten to twenty kilometres in size, and for most of the time they are frozen solid.

Comets are left-overs from the formation of the solar system, about 5000 million years ago. Near to the Sun small, rocky planets such as the Earth formed, further out large gas planets like Jupiter formed, and much further out the masses of dust and ice didn't form planets, they stayed separate and formed a cloud of comets known as

9 This photograph of comet IRAS-Iraki-Alcock was taken on 12 May 1983. By guiding on the nucleus of the comet the photographer has been able to show much of the detail in the coma region of the comet. PHOTO: KEN WALLACE

Oort's cloud. Scientists estimate that this cloud reaches out half way to α Centauri. The comets mostly stay out at that distance, completely undetectable from the Earth, but occasionally two of these objects may collide, or one may be forced out by some other means. When this happens one of these icy pieces may fall in towards the Sun. The journey it is on will take many hundreds of thousands of years and once it reaches the Sun it will swing past it and back out to Oort's cloud, returning in perhaps a million years time. This is the life of a typical comet, but some comets orbit the Sun much more often than this. These short-period comets have been captured by the solar system.

When a comet travels in towards the Sun it must pass all the planets in the solar system—some of these are small and have almost no effect on the comet, but if it passes close to one of the large planets, like Jupiter or Saturn, it may be trapped. Just as a deep puddle of water will slow a car as it ploughs through, so the gravitational force of a large planet will slow a comet, leaving it with too little energy to leave the solar system again. A comet that is caught in this way is doomed to spend the rest of its life orbiting the Sun every few years or so. Halley's is one of these periodic comets; it orbits the Sun every 76 years. Comet Encke, the comet with the shortest known period, was captured much more securely—it passes the Sun every 3.3 years.

The temperature of the solar system is fairly low for the most part, so the ices that form comets stay solid, but near to the Sun the temperature rises, and ice and organic materials melt. When a comet gets within this region, its surface warms and organic substances such as methane and ammonia are released, water-ice melts then boils into steam, and very quickly the comet is leaving a trail of gas behind it. As the gas is released small particles of dust are also thrown off from the comet, adding to this trail.

The Sun sends streams of particles—known as the solar wind—throughout the solar system 24 hours a day. This wind interacts with objects in the solar system creating the aurora on Earth and other effects. When the solar wind hits the dust and gas left behind by a comet it blows it away, causing the trail always to point away from the Sun. Seen from Earth it appears as though the comet has a glowing tail of gas behind it.

Another effect of the solar wind is to split the tail into two. The gases and steam are much lighter than the dust so they are pushed away faster, giving two distinct parts to the comet's tail. When seen from Earth the two tails of a bright comet are easily separated. The gas tail, because it is pushed away quickly by the solar wind, is always straight and points directly away from the Sun. The dust tail, because it moves less quickly, trails along the comet's orbit and appears curved from our vantage point. A second difference is in the colours of the tails. The dust tail, because it simply reflects light from the Sun, appears red when seen in colour photographs. The gas tail, on the other hand, glows because the atoms in it are excited by the solar wind; this glow appears blue in photos.

The other main part of a comet is the area around the lump of ice. This is called the coma; the icy core itself is called the nucleus. The coma is formed from the steam and gases being given off by the nucleus as it heats up. Although a comet's nucleus is only a few kilometres across, the gases in the coma can expand to over a million kilometres. This ball of gas reflects light from the Sun back to Earth, giving the appearance of a dense sphere surrounding the nucleus; in reality the gas making up the coma is so thin that on Earth we would call it a very good vacuum.

Every time a comet passes the Sun and forms a coma and tail, it loses some of its matter. The gas and dust given off is lost by the comet forever; there is no way that it can regain it. Each pass of the Sun makes the nucleus a little smaller, so a periodic comet will eventually melt away to nothing. Each time Halley's comet passes the Sun it loses about ten metres of material from its surface, or about half a percent of its matter. Periodic comets therefore have a limited lifetime; they will travel around the Sun a few hundred times before disappearing completely.

Because they have been past the Sun so many times, periodic comets don't have as much volatile material in them as new comets, so they don't put on a very spectacular show. Comet Encke with its short period has faded so much that you need a telescope just to see it at all. The very spectacular comets, the ones that are bright enough to be seen during the day, are the comets that have just entered the solar system from Oort's cloud. Unfortunately they arrive unannounced, taking astronomers by surprise and disappear quickly into the outer reaches of the solar system once more.

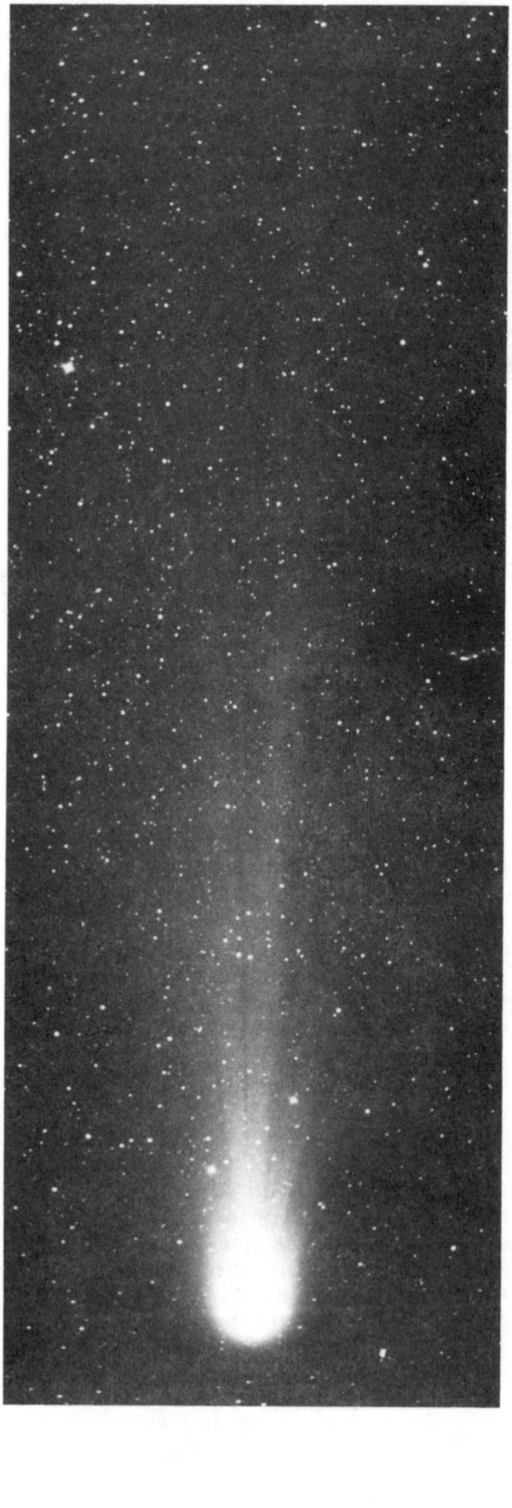

Because of their unpredictable nature, comets are an excellent opportunity for the amateur astronomer to make a contribution to astronomy. Every year many comets come close to the Sun—professional astronomers do not have the time to look out for them all, so quite often a new comet will be discovered by an amateur. Discovering a comet is one of the few ways to be commemorated in astronomy, for if you are the first to sight a new comet, it will be named after you. For example, at the time of writing there have been eleven Comet Bradfields, each one named after Bill Bradfield, an amateur astronomer in South Australia.

It is possible for anyone to discover a comet, but usually it is the result of many hours hard work and dedication. To find a comet you need to have a very good working knowledge of the skies so that anything that is new, or out of place is immediately noticed. You will need to search every night and morning, and you will need a fair bit of luck. Amateur astronomers all around the world vie with each other to find new comets and it may take hundreds or thousands of hours of searching before seeing a new one.

Every time a new comet is detected, astronomers attempt to work out facts about it. From observations taken over a few nights an astronomer can very accurately calculate the comet's orbit and predict where in the sky it will appear over the next few months. These predictions, called the comet's ephemeris, are published quite quickly and your local astronomical society will usually have a copy.

A prediction that is much less accurate is the brightness of the comet and the length of its tail. Despite the extensive research done on comets, we still do not know enough about them to accurately predict these features. A classic example was comet Kohoutek in 1973; predictions were that it would be the comet of the century, but it turned out to be a real flop; it was barely visible to the naked eye. Always take predictions of cometary brightness with a grain of salt, as science still has a fair way to go in some respects.

10 Halley's comet in 1986 showed two tails quite clearly throughout much of its apparition. In this photo you can see the curved dust tail and straight gas tail of the comet. PHOTO: KEN WALLACE

3

Choosing a telescope

Although it is possible to do a lot of astronomy with the naked eye or binoculars, the time will arrive when you want a telescope. This is the time that will determine whether or not you will stay with astronomy, for choosing a correct telescope will ensure many hundreds of hours of enjoyment, while choosing the wrong one will quickly lead to frustration. A quick look at the variety of telescopes available in any specialist shop will convince you that the task of making a correct choice is not a simple one. Before examining the best way to select one, it is necessary to look at the different types of telescopes that are available.

Telescopes divide broadly into two categories, depending upon how they form an image: the refractors and the reflectors. Refractors form an image using a series of lenses, while reflectors, as their name suggests, use a mirror to form an image.

When Lippershey invented the first telescope, it was a refractor. He used a large spectacle lens at one end of a tube, with a second lens at the other. He then looked through the tube and saw a magnified image of a distant object. Although they are now a little more sophisticated, very little has changed with refracting telescopes. Figure 11 is a diagram of a typical astronomical refractor.

Reflecting telescopes are a more recent invention; Isaac Newton made the first one in 1671. By using a specially curved mirror instead of a lens, light from a distant object is brought to a focus so that it can be viewed. Usually at least one small secondary mirror is used to move the focus to a position you can get your eye to. Through the years there have been many modifications to Newton's basic design, resulting in a number of different forms of the reflecting telescope. Figures 12, 13 and 14 show these types, although only two, the Newtonian and the Cassegrain telescopes, are in common use amongst amateurs.

The first decision you must make when you decide you want a telescope is 'How much am I willing to pay?' The least you will have to pay is three or four hundred dollars, although you can spend many thousands if you are able. When looking at a telescope, do not be swayed by the large selection of accessories that come with it. When you start you will need no more than a few eyepieces and you can build up your collection of extras as time goes on if you find that you need them. It is better to start by spending your money on the telescope and mount only, buying the best you can afford. This will give you a good foundation in astronomy, one on which you can build in the future. This is preferable to spending your money on a poorer quality telescope and a box of goodies.

When it comes to choosing between refractors and reflectors there are several points for and against each. Although you must be careful with all precision optical instruments, refracting telescopes with their lenses separated by a sealed metal tube, are much more robust than reflectors. Reflectors are notorious for going out of alignment. Making the necessary adjustments is fairly simple and can be done at home, but you should consider this factor when choosing the style of telescope.

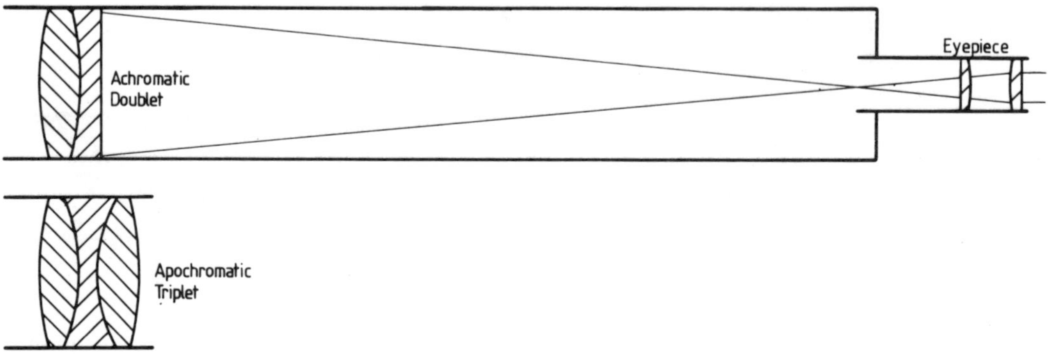

11 The astronomical refracting telescope seen here shows the main parts of these instruments. The achromatic doublet at the front of the scope is sometimes replaced by an apochromatic triplet for better colour correction and fewer aberrations.

Another thing to consider is portability. Once you get started you will probably want to transport your telescope to view from the best sites. If this is important to you, then make sure that your telescope will fit comfortably into your car or van. Normally you will want to carry the telescope inside the car, rather than in the boot, to protect it from bouncing around. Of course, a telescope is much too delicate to carry on a roof rack.

Apart from practical details, there are a number of optical considerations to take into account when selecting an instrument. The most fundamental of these is aperture, the size of the telescope's light collector. The aperture of a telescope is the diameter of its lens or mirror expressed in millimetres, or metres for larger telescopes. This measurement sets the absolute limits for the telescope's performance.

The aperture of a telescope determines both how much detail you can see on astronomical objects, and the faintest magnitude the telescope will pick up. The larger the aperture, the better the instrument is for astronomy. We have seen with binoculars that the bigger the front lenses are, the more light they collect; this is the same for telescopes. When talking of telescopes this is usually expressed as a limiting magnitude, that is, the lowest magnitude star you can see under perfect conditions. The exact limiting magnitude will depend upon the efficiency of the mirrors and lenses, but it is possible to give a guide for most telescopes. Table 3.1 lists the limiting magnitude of some telescopes. It should be noted that these are theoretical limits; effects such as sky brightness or imperfections in the instrument are not taken into account, neither are lens coatings or other improvements.

Another thing that the aperture of the lens or mirror affects is the resolution of the telescope. Table 3.1 lists the resolution of some telescopes. The measurement of resolution will be discussed in chapter 9. For the moment just remember that the better the resolution of the telescope the more detail you will be able to see. This will make the view of the planets and nebulae much more interesting. Resolution also determines

Table 3.1 Limiting magnitudes and resolution of telescopes

aperture	minimum magnitude	resolution
30mm	9.5	3.9″
50mm	10.6	2.3″
60mm	11.0	1.9″
75mm	11.5	1.5″
100mm	12.1	1.2″
150mm	13.0	0.8″
200mm	13.6	0.6″
250mm	14.1	0.5″
320mm	14.6	0.4″
400mm	15.1	0.3″
600mm	16.0	0.2″
1000mm	17.1	0.1″
3900mm[a]	20.0	0.03″
6000mm[b]	21.0	0.02″

Notes: [a] 3.9 metres is the aperture of Australia's largest telescope, the Anglo-Australian Telescope at Siding Spring Mountain in New South Wales.
[b] 6.0 metres is the aperture of the largest telescope in the world. It is at the Special Astrophysical Observatory near Zelenchukskaya in the Soviet Union.

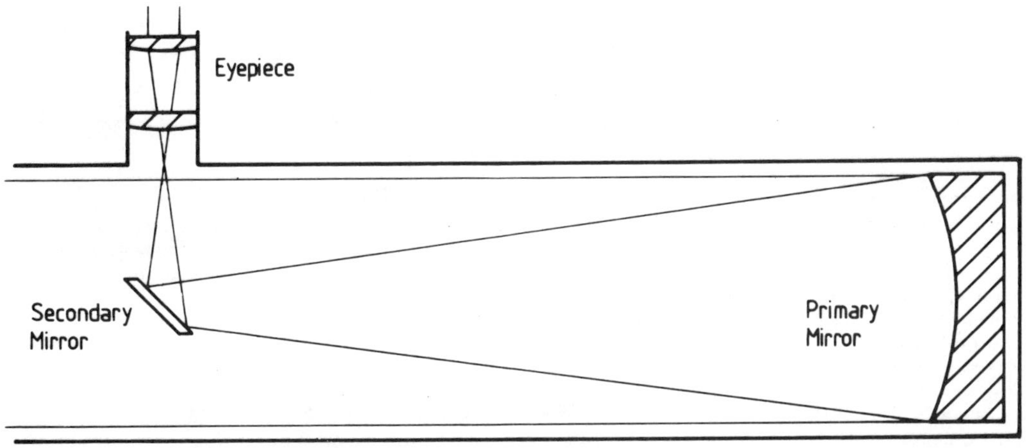

12 A Newtonian reflecting telescope has a paraboloidal primary mirror at the end of a tube with a small, flat secondary mirror held near the top to deflect the focused light into the eyepiece through a hole in the tube.

how close together you can see double stars. Better resolution means a better telescope.

It is obvious that, all else being equal, a telescope with a large aperture is better than one with a smaller aperture. The only problem is that the larger the telescope, the more it costs. In general, a reflecting telescope will have a larger aperture than a refractor with the same price. In a reflector though, you lose a small amount of light because of the obstruction caused by the secondary mirror, though this is a minor effect. There is an aperture for telescopes below which it is useless to go. For refractors this is 50mm, for reflectors it is 100mm, below this a telescope is little better than a toy. Above these apertures there is much good work that can be done. Both Newton and Galileo started off with telescopes smaller than this, 25mm in aperture in fact, but they were pioneers. Today you need a larger instrument to see many of the new discoveries made since their times.

Although the aperture of a telescope is the most important measurement, the focal length also affects the telescope's performance. The focal length is the distance from the mirror or lens to the place that all the light rays come together, the focal point. As we will see in chapter 4, this determines the magnification that an eyepiece gives and also the way the telescope behaves if you use it as a camera lens.

To make things confusing the focal length of a telescope is not usually given in millimetres. It is instead related to the telescope's aperture. This measurement is called the f ratio, and is written f/. To calculate the f ratio you need two measurements, the aperture in millimetres and the focal length also in millimetres. Divide the focal length by the aperture and you have it. For example, a telescope with an aperture of 50mm and a focal length of 600mm has a focal ratio of 600/50 or 12. This would appear on the telescope as f/12.

In general refracting telescopes have an f ratio between about f/8 and f/16 although some new refractors using special lenses go down to about f/5. Reflectors have ratios in the range f/3 to f/10, but some types, especially Cassegrains, can go to f/20. Generally the focal length should not be a major consideration when selecting an instrument, except that it affects the overall length of the telescope, and hence its transportability. Try to avoid telescopes with very short f ratios, however, as that will make it difficult to buy good eyepieces.

Because they are the most common small telescope, I will discuss the selection of a refractor first. There are a number of factors you should use to guide you in your selection of this type of telescope: the first is price. Do not look at the very cheap end of the market, the telescopes offered at 'bargain basement' prices are rarely of any use for astronomy. This is not to say that just because a telescope is expensive it is of good quality; there are some very poor telescopes for which ridiculous amounts are asked. In all cases

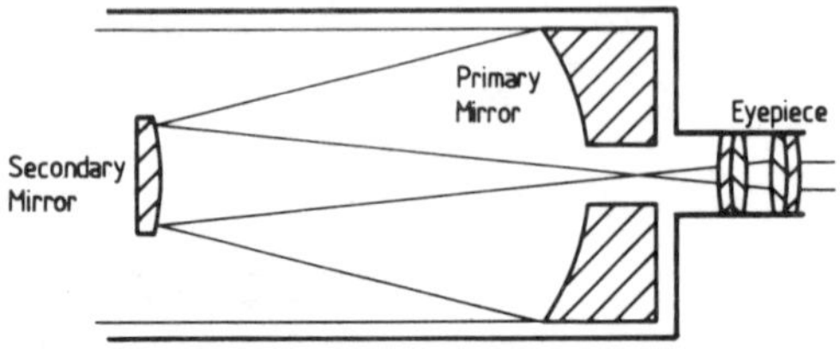

13 The Cassegrain reflecting telescope uses a small hyperboloidal secondary mirror to increase the apparent focal length of the telescope without increasing the overall length of the tube. This allows the long focal length of a refractor to be combined with the short tube of a reflector.

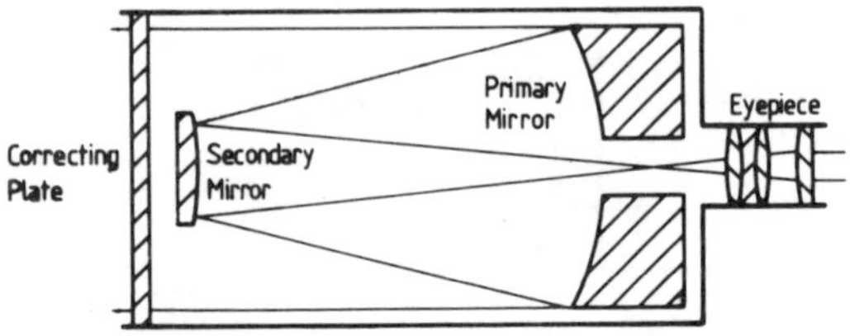

14 The optical layout of the Schmidt Cassegrain reflecting telescope. The correcting plate at the front allows a spherical mirror to be used for the primary. A small hole cut in the middle of the primary mirror allows the light to reach the eyepiece.

it is the optical quality of the instrument that is important.

As it does most of the work, the lens, or object glass, is the part of the telescope that you need to examine most closely. In addition to bringing light to a focus, a lens has some other effects on the light from a star. These other effects, which lower the quality of the image, are called aberrations. Opticians have devised a number of methods of reducing these aberrations to acceptable levels. It is the use of these corrective measures that will determine the cost and quality of a refractor.

Light comes in different colours and a lens affects each colour differently, bringing them to a focus at slightly different points. This is the principle used in a prism to separate the light into its colours. When this happens in a lens each image ends up with a coloured halo around it, an effect called chromatic aberration. To lessen this effect, two lenses, each of a different type of glass, are used as one. This is shown in figure 11. The arrangement is known as an achromatic doublet. Nearly all refracting telescopes have such a doublet, but make sure, as the colour haloes of chromatic aberration make any astronomical work impossible.

A point to check about achromatic doublets is that the two lenses are cemented together. Because we look at very faint objects, every bit of light coming through a telescope is precious and shouldn't be wasted. Unfortunately, every time light strikes a surface, some of it is reflected away. This doesn't only happen at the front lens surface of a telescope, but also at the surface where the two parts of an achromatic doublet join. At this surface as much as 9% of the light can be lost due to reflection; by cementing the two lenses together this can be reduced to 1%. The simplest way of finding out if an object glass is cemented is to look in the telescope's instruction book; cemented doublets are usually listed as such, while non-cemented ones are usually called 'air spaced' or 'simple' doublets.

Another way of making a telescope lens achromatic is to use three lenses instead of just two. This system gives almost perfect colour correction and much better image quality. Lenses made using this system are called apochromats. Apochromats are much more expensive than normal achromatic lenses but they considerably enhance a telescope's performance. Some of the new high quality refractors are being fitted with apochromats. They are a worthwhile extra expense if you can afford it.

Another cause of light loss is the reflection at the front of the telescope's object glass. This is usually about 5% but, as we have seen for binoculars, it can be reduced by coating the lens surfaces with thin layers of transparent materials. These coatings can reduce the amount of lost light to as little as 0.1% at each surface, increasing the clarity of the view enormously. Another thing a lens coating will virtually eliminate is ghosting. Ghosts are faint images of bright objects, which appear to one side of the object, and which are caused by reflections of light from uncoated surfaces. They can be very annoying when you are looking for faint objects. Coated lenses are usually called such in the literature that comes with a telescope and they are a worthwhile extra expense.

It is very difficult to check to see that a telescope lens is of good quality when looking at it in a shop, but there is one way that you can tell if it has any major faults. To remove any

15 Amateur astronomer Steve Janesse demonstrates the use of an equatorially mounted refracting telescope. PHOTO: DAVID REIDY

aberrations caused by a poor lens, manufacturers mask part of it. Put your eye to the eyepiece end of the telescope, without an eyepiece in place, and look at the object glass. You should be able to see all the lens at once, there should be no rings or baffles obscuring your view of any part. If there is something masking the edges of the lens do not buy the telescope. Even if there is nothing wrong with the lens itself, the extra blockages will lessen the amount of light that reaches your eye.

When choosing a reflecting telescope it is the mirror that you need to inspect. Because all light reflects the same way there is no chromatic aberration from a mirror, nor is there any problem from extraneous reflections. When inspecting a mirror it is necessary to find out if it has the correct shape. The shape of the mirror must be correct to a very high accuracy, within a forty thousandth of a millimetre in fact. To avoid using too many small numbers, opticians measure the accuracy of a surface by comparing it with a wavelength of light. This is written as $\lambda/8$ or $\lambda/10$ for an accuracy of one eighth or one tenth of a wavelength of light, about 40nm. To be suitable for astronomy a mirror must be accurate to within one-eighth of a wavelength, or better. This information is usually included in the instructions that come with a telescope. The best mirrors are $\lambda/10$ and you will not find any marked better, as no optician's instruments can measure them any more accurately.

Telescope mirrors are coated on the front surface of the glass rather than on the back as is usual with most domestic mirrors. It is this thin coating of metal that does the work of making an image. There are two types of coatings that are used on glass mirrors, aluminium and silver. Silver has been used almost since the invention of the reflecting telescope, but it is now fairly uncommon as aluminium is much more durable. It would be fairly rare to find a silvered telescope, but check for this when purchasing. When choosing a telescope make sure that the mirror has an even and smooth coating of metal, and that it has no dull spots.

The problem with using metal films in telescopes is that they tarnish, silver much more so than aluminium. A freshly coated silver mirror will reflect about 93% of the light falling on it, but as it ages this will fall to about 77%. Unfortunately, the pollution in the air of our cities will hasten the tarnishing of silver. A new silver surface will only last three to six months before it needs replacing. Aluminium coatings, though they reflect only 88% of the light falling on them, last much longer, usually five to ten years. Sodium chloride (common salt) is one of the few common chemicals that affects an aluminium film, so living close to the sea will shorten its life. Some high quality mirrors have an additional coating of protective material to increase the lifetime of a mirror. This is useful, but not essential as it is fairly simple and inexpensive to get a mirror re-aluminised.

If you have either a silvered or aluminised mirror that needs to be re-coated, take it to a telescope shop or back to the manufacturer who will arrange it. It is possible to re-silver a mirror yourself, but it is a complicated process and involves toxic and dangerous chemicals and is not worth the trouble. Because of the expensive vacuum equipment needed to coat glass with aluminium, you can't re-aluminise a mirror at home.

Although the mirror or lens of the telescope is the most important part, there are a number of other points to bear in mind when selecting an instrument. The outer finish of a telescope should be unscratched and not dented. Although it may seem a minor point, signs of rough handling on the outside of the telescope may mean that the

16 The primary mirror of a Newtonian reflecting telescope is here seen in its mirror cell. PHOTO: DAVID REIDY

optics have been put out of alignment. Fixing this is a major task for a refracting telescope.

After the telescope itself, the other piece of equipment you will need to purchase before you can do any astronomy is a mount to stand it on. There are two types of mounts to choose from: equatorial mounts and altazimuth mounts. Each type works on a similar principle, but the effect of each is very different. Equatorial mounts, while much more expensive than altazimuth mounts, are the only sort you can use for photographic work and for locating objects easily.

As you know, the stars and planets rise in the east and set in the west throughout the night. This appears as if the whole sky is rotating about two points in the sky, the celestial poles. The north celestial pole, NCP, has the fairly bright star, Polaris, close to it, while the south celestial pole, SCP, the one visible from the southern hemisphere, has no such star at the moment. This motion of the stars appears through a telescope as the stars moving in one side of the view and out the other. This motion is fairly slow and you can move the telescope to compensate for it if you are simply observing, but if you want to do photography it can be a problem.

Because of the apparent rotation of the sky about the SCP a telescope which is mounted on a mount which has one of its axes pointing at the pole can follow the stars by moving only about that axis. This is the principle of the equatorial mount. The main axis, the polar axis, points at the SCP. The other axis, the declination axis, is at right angles to the polar axis and is used to move the telescope to look at different parts of the sky. Though there are a number of systems of equatorial mounts, the one that is used most commonly in amateur telescopes is the German type. This form of the mount has the telescope held in a ring or cradle on one side of the polar axis and a counterweight on the other side to balance it. Figure 18 shows this arrangement. Setting up an equatorial mount is not a simple task and we shall see how to do this in chapter 6.

If you are buying a Newtonian telescope with an equatorial mount there is another point to look for. When you are using the telescope, it is possible to end up with the eyepiece in some strange places and often it is difficult to get your eye to it. A mount that will let you rotate the telescope within its cradle without moving the rest of the mount is most useful. This will allow you to turn the telescope so that the eyepiece is back in a convenient position. Although this feature is not essential, it will make viewing much easier.

The other sort of mount is much more common; it is the type used for camera tripods. It is called the altazimuth mount. The name comes from the fact that the mount allows the telescope to move in two directions, up and down, altitude, and from side to side, azimuth. By combining these two movements the telescope can be made to point at any part of the sky. The disadvantage of this system is that you must move it in both axes simultaneously to follow any object through the sky. It is, however, much simpler to set up than an equatorial mount.

When choosing a mount of either type the most important thing about it is that it be steady. A mount that cannot support the weight of a telescope and hold it securely against the wind is useless. The easiest way to test the strength of a mount is to put the telescope on it, focus it on a distant object and, while you are looking through it, give the mount a tap with your foot. The tap should be firm, but not enough to move the legs of the mount. If the mount is strong enough for the telescope, the view through the eyepiece should return to being steady after a second or two. If the view doesn't stop shaking in this time, or it has moved from the object, then the mount won't be firm enough to support the telescope in a breeze.

Another point to check is that the mount

17 The simple altazimuth mount on this telescope is one method of mounting a telescope for astronomical use. The adjusting brace on the side of the mount allows small adjustments in altitude to be made while viewing.
PHOTO: DAVID REIDY

moves smoothly. While looking through the eyepiece use the controls on the mount to move the telescope. The view should move smoothly, and there should be no backlash at the end of the motion. This extra motion, although it will not affect the mount overall, can be annoying. Some telescopes have very small mounts designed to sit on top of a table or other support. These are fine for terrestrial use, but they are quite unsuitable for astronomy. You may have a table in your backyard that you can use to view, but if you take your telescope out to the country to observe you will probably find yourself lying in the dirt to get your eye to the eyepiece.

When Halley's comet was approaching many people bought telescopes to see the phenomenon. Now that it has gone there are a number of hardly used second-hand telescopes becoming available. One advantage a new telescope has over a used one is that before it leaves the workshop for the showroom, most manufacturers check their products to ensure they are working correctly. Another reason for buying a new telescope is that manufacturers must guarantee their products; this gives you added protection if the instrument is faulty. Taking all this into account though, there is no reason why you shouldn't save some money by purchasing a second-hand telescope, provided you take a little more care when checking it over.

When checking the lens of a second-hand telescope, take the same care you would with a new one: check for marks and scratches and that the object glass is achromatic and cemented. To check that an object glass has an anti-reflective coating examine the reflections you get from the surface. If there is a coating, the reflections will be very faint and will have a purple colour. A second check is to shake the telescope; do this gently and listen for anything loose. It may just be a screw or bolt, but if the sound comes from the main lens forget about the telescope. A loose object glass will need to be re-aligned by an optician.

When looking at a used reflecting telescope examine the mirror for scratches, or an uneven coating as these will make the telescope unsuitable. A mirror that has a tarnished coating will need to be re-aluminised. This can easily be done at a telescope shop, and shouldn't deter you from buying the telescope, but ask for a discount. Sometimes mirrors will have small nicks out of them, especially around the edge. As long as these are not too large or too numerous they will not noticeably affect the mirror's performance. It is best to paint these nicks out with matt black paint before using the telescope; again, ask for a reduction in price.

In addition to the normal checks you should make on the mount, examine it carefully for deterioration. Wooden tripods are sometimes attacked by borers; these leave small holes in the legs and will make a mount unsuitable for use. Metal tripods sometimes rust, as they are often out in the damp air. Check for any rust that is more than just a surface coating. Remove any rust when you buy the telescope and paint the affected areas with rust inhibitor.

Finally, put a low powered eyepiece in the telescope and examine the view of a distant

18 The home-made reflecting telescope in this picture is mounted on a simple motorised equatorial mount. Note the three massive counterweights on the end of the declination axis needed to balance the telescope.
PHOTO: DAVID REIDY

object, not the Sun, through it. The image should be bright and clear. If possible try to use the telescope at night, looking at the stars. The stars should appear as small round points of light all through the field; they should have no coloured haloes. With a reflector the stars may appear to have small spikes. This is not a fault but an effect caused by the holder for the secondary mirror. For some types of observing the spikes are an advantage. If possible, take someone who knows a little about astronomy to also look through the telescope, or your local astronomical society may be able to help you.

4

Accessories for your telescope

Once you have selected a telescope, you will find that there are hundreds of different accessories you can purchase for it. Some of these are necessary, many are useful, and a few are a waste of money. Because of the wide range of extras, I feel that it is better to start observing before you spend, work out what you need and then buy it. This way you won't be left with pieces you never use.

The first accessories you will have to buy for your telescope are eyepieces. These are systems of small lenses that fit into the end of the telescope. They magnify the image from the lens or mirror so that you can see it. Without an eyepiece, your telescope is useless as it is almost impossible to view the image made by the main mirror or object glass with the naked eye.

Eyepieces come in all shapes and sizes and with all sorts of lens arrangements. Choosing among them can be difficult. The most important feature of an eyepiece is its focal length, for it is this that determines the magnification that your telescope will give. The focal length of an eyepiece is recorded on the eyepiece somewhere, and is always given in millimetres. To calculate the magnification that a given eyepiece will give with your telescope you will need to know the telescope's focal length in millimetres as well as the focal length of the eyepiece. The magnification is given by dividing the telescope's focal length by that of the eyepiece. For example, a telescope with a 600mm focal length will give a magnification of 50× when used with an eyepiece with focal length 12mm (600/12 = 50). The small × after the 50 is the symbol for magnification. It is pronounced 'times', so 50× is 'fifty times'. Table 4.1 shows how to determine the magnification a given eyepiece will give with a given telescope.

Because they determine the magnification of the telescope, you will probably want a range of eyepiece sizes, but choose carefully. It may seem at first glance that you should buy the eyepieces with the shortest focal lengths as these will give the greatest magnification. A 2mm eyepiece in a 600mm telescope will give 300×, but there are problems. A telescope has a maximum magnification beyond which, although you can buy the eyepieces, it is useless to go. A telescope can

Table 4.1 Magnifications given by eyepieces and telescopes

		telescope focal length (mm)									
		500	600	700	800	900	1000	1200	1500	1750	2000
eyepiece focal length (mm)	4	125	150	175	200	225	250	300	375	438	500
	6	83	100	117	133	150	167	200	250	292	333
	8	62	75	88	100	112	125	150	188	219	250
	10	50	60	70	80	90	100	120	150	175	200
	12	42	50	58	67	75	83	100	125	146	167
	15	33	40	47	53	60	67	80	100	117	133
	20	25	30	35	40	45	50	60	75	88	100
	25	20	24	28	32	36	40	48	60	70	80
	30	17	20	23	27	30	33	40	50	58	67
	35	14	17	20	23	26	29	34	43	50	57
	40	12	15	18	20	22	25	30	38	44	50
	45	11	13	16	18	20	22	27	33	39	44
	50	10	12	14	16	18	20	24	30	35	40

Note: To use this table, firstly find the column that corresponds to the focal length of your telescope, then look at the row for the eyepiece's focal length. You can read off the resulting magnification quite easily.

19 A set of 1.25″ eyepieces ranging in focal length from 6mm to 40mm. PHOTO: DAVID REIDY

only see so much, and if you try to magnify beyond its limit you will simply get a poor quality image. To calculate the maximum magnification for your telescope you multiply its aperture in millimetres by two. For example, a telescope with a 50mm aperture has a maximum useful magnification of 100×.

Just because there is a maximum magnification doesn't mean that you can use it all of the time. The calculated maximum is a theoretical limit, in practice you can rarely use this much power. The unsteadiness of the air at night and imperfections in the telescope or its mount, are all factors that will affect the actual maximum magnification you should use on any night. Of these, it is the unsteadiness of the air that is the biggest problem. You are probably familiar with the way stars twinkle on some nights; this is caused by currents of air in the upper atmosphere. Because of the magnification of a telescope, this effect is enlarged and the stars seem to be jumping all over the place. Reducing the magnification reduces the effect so that you can continue viewing.

There is also a lower limit to the magnification you can use. It is calculated by dividing the aperture in millimetres by eight. The 50mm telescope above has a minimum magnification of about six (50/8 = 6.3). If you try to use a lower magnification than the minimum you will be wasting some of the light that the telescope collects. You should calculate the maximum and minimum magnifications for your telescope and bear these in mind when you are choosing eyepieces. Three eyepieces are usually enough to give you a good start: choose one that is about 75% of the maximum magnification, one that is about twice the minimum magnification and a third about halfway between these two.

After magnification, the most important feature of an eyepiece is its field of view. This describes the size of the circle you see when you look through the eyepiece. The larger the field of view of the eyepiece the better the image you will see. A large field of view is more fun than a narrow one as it removes the feeling of looking down a long pipe. Some eyepieces have fields of view as large as 70°, while 40° to 50° is quite good.

Finally consider eye relief. This describes the distance from the eyepiece that you need to hold your eye to view. This is important as it affects your comfort; holding your eye very close to an eyepiece for any length of time is tiring. Eyepieces with large eye relief that allow you to hold your head back a little are much better, especially if you want to show your telescope to people who are not used to using one. They will find these eyepieces much easier to use. If you wear spectacles, some large eye relief lenses will allow you to use the telescope without removing them, a big advantage. Generally low power eyepieces have better eye relief than high powered ones.

Make sure that the eyepieces you are looking at will fit your telescope. There are two main sizes of eyepieces. The smaller size has an outer diameter of 24.5mm; this is common in telescopes from Asian manufacturers. The second standard is a 31.8mm outer diameter; this is more

20 Four eyepieces of differing focal lengths and diameters show the range of sizes available to the amateur. PHOTO: DAVID REIDY

commonly known as a 1¼ inch eyepiece. It is the standard for American and European telescopes, and is also used in most large amateur telescopes. At one time 50.8mm (two inch) eyepieces were used, but these are rarely seen nowadays.

Unfortunately selecting an eyepiece is not as simple as just picking a focal length; there are many different styles to choose from. Each of the different forms of eyepiece has particular strengths and weaknesses, but there are a few points that are common to all. Always pick coated eyepieces; these have the same anti-reflective coatings as object glasses and binoculars and generally give a view superior to uncoated optics. Always give a new eyepiece a shake, there should be nothing loose in it. If there does appear to be a rattle reject the eyepiece as having them fixed is quite expensive. There are a huge number of different designs for eyepieces; those below are some of the ones commonly used for amateur astronomy.

One of the earliest types of eyepieces invented was the Ramsden. These have a field of view of 35°, but suffer from chromatic aberration as they have only two lenses in them. In addition to these drawbacks they also have poor eye relief. The only point in favour of the Ramsden eyepiece is that it is non-cemented making it suitable for solar work, as we will see in chapter 8.

Two improvements on the Ramsden eyepiece are the Kellner and the achromatic Ramsden. The Kellner is a semi-achromatic eyepiece that has a wide flat field of about 45°. It has better colour correction than the Ramsden, but it is haunted. That is to say that it has a large number of ghost images in it. The achromatic Ramsden is similar to the Kellner except that it has full colour correction. Both eyepieces have little eye relief, but they can usually be bought quite cheaply.

The last of the early types of eyepieces is the Huyghenian. This eyepiece is commonly included with small telescopes. It is usually not achromatic although it has some colour correction in its design. It has very little eye relief and only a 20° field of view. Another problem is that it cannot be used successfully with telescopes of f ratios less than f/12.

The 'orthoscopic' eyepiece was invented to correct many of the problems of earlier eyepieces. By using two pairs of cemented lenses, the orthoscopic is achromatic. It has a fairly wide field of view and moderate eye relief. Orthoscopics are relatively inexpensive and a set of these will give you a good start for general observing.

In recent years a number of complicated eyepiece designs have become available. These eyepieces, although they are more expensive than the earlier types, usually combine good colour correction, wide fields and reasonable eye relief.

The Plossl eyepiece is available from a number of manufacturers. It uses a pair of achromatic doublet lenses to give a colour corrected, 50° field. The eye relief is about 80% of the focal length, so a 25mm eyepiece has 20mm of eye relief.

By using three sets of lenses instead of two, opticians were able to design some spectacular eyepieces. The most common of these is the Erfle. Using three achromatic doublets, these eyepieces have a 70° field of view and an eye relief of 50% of the focal length. When choosing an Erfle eyepiece, it is most important that it has coated lenses. If it doesn't, the ghosting and light loss will make it very difficult to use.

When you have decided on a type of eyepiece you will find that there is an advantage to buying all the same brand. Some manufacturers match their ranges of high quality eyepieces so that they

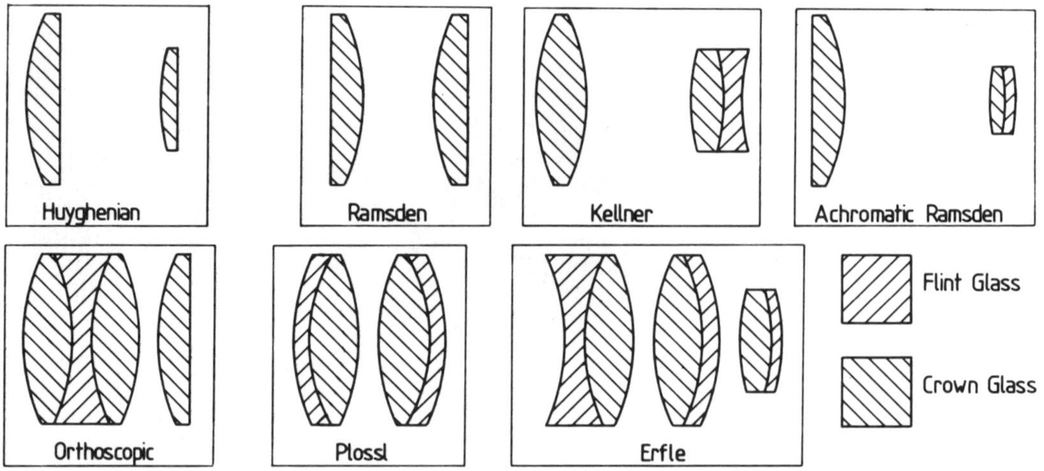

21 The optical arrangements of seven of the most popular eyepieces are seen here. Those at the top are the early forms of eyepiece, those at the bottom are more modern forms. Two types of glass are used in eyepieces—crown glass and flint glass. Each has slightly different optical properties allowing opticians to remove chromatic and other aberrations by carefully combining lenses of each type.

all focus at the same point. This means that you can change from one eyepiece to another without changing the focus of your telescope. Having these parfocal eyepieces is an advantage, especially if you like to change magnifications frequently.

One very common accessory for eyepieces is the Barlow lens. This lens modifies the light from the telescope before it enters the eyepiece so that the eyepiece gives two or three times its normal magnification. The Barlow is a quick and easy way to double the range of magnifications you get from your eyepieces. It will cost you about the same as an eyepiece so it is also a cheap alternative to a large range of focal lengths. Unfortunately there are also some drawbacks to counter all these advantages.

Barlow lenses usually introduce some amount of chromatic aberration and ghosting to the image. These can be lessened by using coated and achromatic Barlow lenses, but they will still be present to a small degree. When selecting a Barlow bear in mind that although it increases the magnification of your telescope, the upper limit on magnification doesn't change. Eyepieces that already give you more than half the maximum magnification for your telescope will not be of any use with a 2× Barlow.

When using a Barlow it is fitted into the eyepiece holder, then the eyepieces are fitted into the Barlow. This may be a problem with some small Newtonian telescopes as a Barlow that fits a long way into the eyepiece holder may foul the telescope's secondary mirror. Finally, Barlow lenses are usually too small to capture all the light from the mirror or object glass, so images are generally fainter than with an eyepiece of the same total magnification.

Another accessory, without which it is almost impossible to use a telescope properly, is a finder scope. This is a small telescope or other device mounted on the side of the main telescope to help you point the instrument in the correct direction. This will allow you to find the objects you would like to see. Without such a help, you will find it impossible to point the telescope accurately. Some telescopes have an in-built finder that works using a small prism which flips down into the main telescope. This then directs some of the light up into a special eyepiece. Although this seems a good idea, it is almost useless. The prism is too small to make the image bright enough to see anything well. If your telescope is fitted with such a device, you will still need to buy a separate finder.

Some telescopes, instead of providing a finder, have a system of sights similar to those on a gun. This system, though very simple, can be used quite successfully for easy pointing tasks. It should suffice to get you started in astronomy and

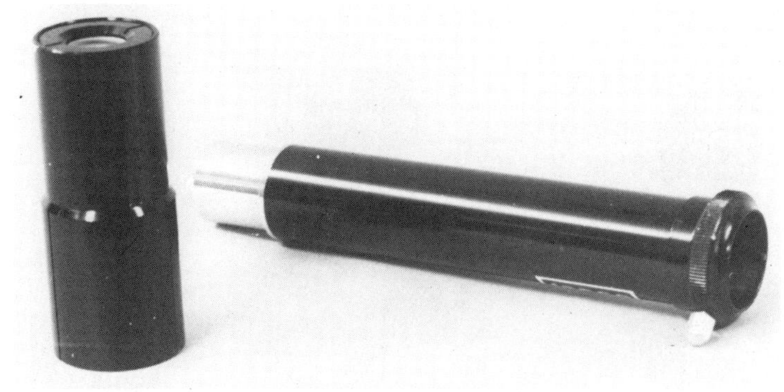

22 Two 2× Barlow lenses. Using lenses such as these will double the range of magnifications of your eyepiece collection. PHOTO:DAVID REIDY

you can buy a proper finder scope later when the need arises.

When choosing a finder scope look for one with a reasonable aperture and a low magnification. Finders are labelled in the same way as binoculars, by their aperture and magnification. Of course a finder with coated achromatic optics is better than one without, though this is not as important as it is for your main instrument. Finders start at an aperture of about 25mm and range up to the size of many main instruments. One of the finders on the 3.9m Anglo-Australian Telescope has an aperture of 60cm, bigger than all but the largest amateur instruments. If you are only going to do casual observing a small finder with magnification of five or six times is satisfactory. If you want to do photography, then you will need a 30mm or 40mm finder at least.

The eyepieces in finders are exactly the same as those in telescopes, but they usually have a set of crosshairs inserted in them. Make sure that the crosshairs are in focus when you look through the finder, or at least adjustable so that you can focus them. Out of focus crosshairs will make it difficult for you to use the finder effectively and may well give you a headache after using them for a while. Some expensive finders have the ability to illuminate the crosshairs. This is useful if you intend to do any photography, but there is usually enough sky brightness to make non-illuminated crosshairs visible enough for most finding tasks. If you do invest in an illuminated finder, make sure that you can adjust the brightness to suit yourself. The amount of lighting you need will vary from night to night according to seeing conditions.

These small finders come in two styles, straight-through and right-angle. The straight-through finders, as the name suggests, have their eyepieces in a straight line with their object glasses. The right-angle finders have a small prism to turn the light through 90° to make it easier to get your eye to the eyepiece. I prefer the straight-through type of finder, though other astronomers I know will only use the right-angle type. The simple system for pointing the telescope I will describe in chapter 6 will only work if you have a straight-through type. You can buy finders that adjust so you can use them straight-through or at right angles. It may be better to try both types before you buy so that you can get the type that suits you best.

Finders are mounted on the side of a telescope using one or two mounting points. Exactly how these mountings work varies from manufacturer to manufacturer, but the important point is that the finder must be easily adjustable so that it points in the same direction as the main scope. The mounting system must be firm enough to hold the finder steady once it is aligned, and you should be able to lock the setting when it is correct. In chapter 6 I will tell you how to align the finder properly, but when you're buying just check that the finder doesn't move if you try to shake it gently.

An accessory that is particularly useful on refracting and Cassegrain telescopes is the star diagonal. This is a device that fits into the eyepiece holder and which turns the light through a 90° angle. When you are viewing close to the vertical with a refracting telescope, you must twist your head so that you can look upwards into the eyepiece. With a star diagonal you can adjust its position so that you can look downwards. It makes it a lot easier to get your eye to the eyepiece. Newtonian telescopes, because they

23 A simple, straight-through finder scope in a single mounting bracket. A finder such as this will allow you easily to locate objects to view through your main scope.
PHOTO: DAVID REIDY

have the light emerging from the side, do not suffer from the problem if they are on a mount which allows the telescope to be rotated.

Star diagonals come in two forms, ones that use mirrors and ones that use prisms. Mirror star diagonals need to be cared for in the same way as telescope mirrors and they will need to be re-aluminised every few years. Prism diagonals work on the principle of total internal reflection so they have no reflective surface to tarnish. Another advantage of prism diagonals is that they cannot go out of alignment, something that can happen to mirror types. When choosing a prism diagonal, check that it is coated and that it is not loose inside its holder.

Another form of attachment that goes by the name of diagonal is the sun diagonal or Herschel wedge. This is similar to the star diagonal, except that instead of using a prism to turn the light, it has a flat piece of ordinary glass. Using glass lets most of the light from the Sun pass out of the telescope leaving only a little for the eyepiece. These devices are very dangerous and should never be used. The Sun is so bright that there is only one safe way to view it as we will see in chapter 8. The main problem with the Herschel wedge is that the extra light and heat needs to go somewhere, usually this is onto your neck or clothing causing severe burns. If you want to view the Sun use the safe eyepiece projection method described later.

Many small telescopes are made to serve a dual purpose, an astronomical instrument at night and a general purpose viewing system during the day. There is no reason why your telescope can't be used for both activities, except for the fact that astronomical telescopes turn everything upside down. To make a telescope give an image that is the right way around requires extra lenses; these lenses will waste some light. As it doesn't matter which way around you see the stars and planets, astronomical telescopes are not fitted with these lenses. In fact, many early maps of the Moon were drawn upside down just for astronomers. Luckily the extra lenses needed for an upright image are available as a telescope accessory—they are known as inverters. The extra light they waste is not a problem during the day as images are bright enough anyway, but don't use them at night as they don't serve any useful purpose on the stars.

Another accessory which is sometimes included with telescope packages is a multiple eyepiece holder. This fits into the telescope's drawtube, then the eyepieces fit into it. This allows you to change the magnification you are using simply by turning the eyepiece holder around. The problem with such devices is that they waste some of the light from the telescope, something you should try to avoid. You will only occasionally need to change the magnification you are using, and it is a simple task to remove one eyepiece from the telescope and replace it with another, especially if you have all your eyepieces arranged ready for use.

In chapter II we will look at a simple method to take astronomical photographs—to do this you will need a camera adaptor of some sort. Camera adaptors come in two types, the ones which allow you to photograph through the eyepiece of the telescope, and those that simply attach the camera to the telescope so that it points in the same direction. When buying such an accessory, make sure that it is strong enough to support your camera securely. If you would like to do some photography, read chapter 11 to decide exactly what sort of adaptor you will need. If you are purchasing an eyepiece adaptor you will also

24 A high quality right-angled finder with double mounting brackets. PHOTO: DAVID REIDY

need a special adaptor for your camera. The eyepiece adaptors end in a standard mounting called a T-ring; for this to fit your camera you will need a T-ring adaptor; these are available for all popular makes of camera and you can probably get one at your local camera shop.

One of the silliest accessories for small telescopes is the coloured filter set. This is usually a selection of red, blue, yellow, orange and green filters which are made to screw into the back end of your eyepieces. In all my experience in amateur astronomy I have never seen the need for a coloured filter of any kind to be used on a small telescope. All a coloured filter will do is to make the image dimmer. There are few astronomical objects which are bright enough to withstand such a waste of light. In the 1960s there was some talk of an effect called a transient lunar phenomenon (or TLP). These were meant to be more visible with some coloured filters, but in recent years the idea of TLPs has fallen into disrepute.

A recent invention has been the nebula filter. This device, also known as a 'light pollution rejection filter', or LPR, allows you to view from a town or city without the annoying effects of light pollution. Using such a filter will make the sky appear very dark, both with the naked eye and with a camera. Unfortunately, the LPR filter will also make everything else look dimmer. Unless light pollution makes observing almost impossible from your location, you will probably not need one of these devices.

Polarizing filters, like coloured filters, are not particularly useful in amateur telescopes. A polarizing filter will do little other than reduce the amount of light reaching your eye. Reflecting telescopes are particularly unsuitable for polarizing filters as the light coming from a mirror is often already polarized by the way the aluminium was deposited on the glass. There are some specialised astronomical tasks for which polarized light is required, but these tasks require a large telescope.

By far the most dangerous accessory, and one that is often supplied with small telescopes, is the solar filter. This is a very dark filter which fits into your eyepiece and is supposed to reduce the level of light enough to allow you to view the Sun. The best thing to do with one of these filters is to wrap it in aluminium foil to keep it safe, and bury it in a deep hole in your back garden. Under no circumstances should you ever use it for astronomy. The amount of heat put into your glass filter from even a 50mm telescope is enough to shatter it. Were this to happen the Sun's rays would enter your eye and blind you instantly. Don't take the risk.

There is a special class of solar filters which are safe: these are the hydrogen-α filters. They work

25 A prism type star diagonal for use with 1.25″ eyepieces. PHOTO: DAVID REIDY

using a very sophisticated system of coatings on a special plate to allow only a minute fraction of the Sun's light through. These filters will allow you to view all the features of the Sun in safety. Unfortunately a hydrogen-α filter costs thousands of dollars. A five dollar sun filter won't keep your eyesight safe.

Moon filters are another type of filter that has become available. These are neutral density filters: that means they don't give any colour to the image; they are used simply to make it less bright. If you have trouble viewing the Moon because of its brightness, it might be useful to have such a filter. Remember though, leave the Moon until last each night as viewing it, even with a moon filter, will ruin your dark adaptation.

Once you have bought a range of eyepieces and accessories you will have the problem of storing them while you are observing. Your pockets are very useful, but unfortunately they are dusty and shouldn't be used, unless you want to be forever cleaning your equipment. A better idea is to buy or build a tray that will attach to your mount and in which you can hold all the bits and pieces. It is possible to purchase such a holder with holes in it to keep your eyepieces in order. It is difficult to read the small markings on an eyepiece in the dark so there is much to be said for keeping all your eyepieces arranged so that you can easily get the magnification you want.

One final accessory is not for your telescope, but for your equatorial mount—the telescope drive. This is a device that you fit to your mount to make the telescope follow an object through the sky automatically. You can observe quite successfully for many years without a drive mechanism, but if you want to do much photography you will need one fairly soon. Usually there is not much choice in the type of drive you can buy, as most manufacturers only make one to fit their mounts.

If you do have a choice, there are a number of features to look for. Firstly, the power source. If you will be observing only from home, a drive that plugs into a power point is fine, but if you're observing in the bush this is a problem. Some drives will run from a small battery pack. These are more expensive to run, but are quite convenient for country observing. Drives that combine both of these methods are the best.

Secondly you should look for the drive that allows you to vary the rate at which it works. Most drives are designed to turn the telescope around completely once every sidereal day. If your telescope is perfectly aligned, that is all you will need. If on the other hand you haven't set the mount up exactly right, you will find that the telescope will slowly drift off the object being viewed. When this happens you will need to make adjustments. Some drives let you make these adjustments in both axes of the mount simply by pressing a button. Such a system is virtually a necessity for most photography through the telescope.

Even if you can get no more than a simple one-speed, no-control drive for your telescope, all is not lost. Another device, called a drive corrector, can be fitted to make your mount do these minor adjustments. Drive correctors often give you an additional feature. They will plug into your car's battery and give you the 240V needed to run the telescope motor. These devices can be bought over the counter from most telescope shops, or you can make your own. If you have a little talent in electronics, your local astronomical society can probably provide you with the instructions to build your own.

26 A selection of 1.25″ and 24.5mm observing filters. The filters screw into the base of your eyepieces to modify the light passing through them. PHOTO: DAVID REIDY

27 The worm gear drive on a German equatorial mount. PHOTO: DAVID REIDY

5

Astronomical terms and systems

Like many other fields of science, astronomy has developed specialised words and concepts to help describe objects and ideas. This jargon is unavoidable and before you advance far in amateur astronomy you will need to understand at least some of it. In this chapter I will describe some of these terms, but it is by no means necessary that you assimilate all that is here in one go. I have included everything in one chapter for the sake of completeness. You can skip a lot of the information presented here and still get started with observing. As you become more familiar with astronomy and talk with other amateurs, the concepts will be clearer and this chapter more useful.

In chapter 1 we saw the magnitude scale used for describing and measuring star brightness. Because the steps in the magnitude scale are so large, a factor of 2.51 times, astronomers divided the scale into smaller segments. The normal magnitude scale is made more exact by having tenths and hundredths of a magnitude step. Using tenths the step size is only 1.10 times, using hundredths each step is only 1.01 times, a much finer scale. Because of this further division, magnitudes are usually written with two decimal places. For example, the star β Crucis has a magnitude of 1.25.

To measure the magnitude of a star astronomers use two methods. The first involves the use of a special instrument called a photometer—magnitudes measured in this manner are known as photometric magnitudes. The second method uses special photographic plates. Kodak make a range of films specifically designed for use by astronomers. These are used to take photographs of areas of sky, then the magnitudes of the stars are calculated by measuring the size of the dot on the film. Magnitudes obtained using this method are known as photographic magnitudes.

As we will see in chapter 9 the colour of a star can give us important information about its temperature, size and composition. Colour is a very subjective thing, one person may well describe the colour of paint differently from another, let alone the very faint light coming from a star. So that the colours of astronomical objects can be determined precisely, astronomers have a system known as colour magnitudes and colour indices. The colour magnitude of a star is determined photographically, but instead of using a normal film, astronomers use special Kodak films and coloured filters to take the photograph. The colours of stars are commonly measured in ultra-violet, blue, red and infra-red light. These are known as U, B, R and I magnitudes. To distinguish normal magnitudes from these coloured ones, they are known as visual, or V, magnitudes.

To make it easier to list colours, a system of colour indices has been developed. These are calculated by comparing the magnitudes of a star recorded in two different colours. The most common colour index is the B-V type. Pronounced 'B minus V', this is the blue magnitude of an object minus its visual magnitude. For example, the star β Crucis has a blue magnitude of 1.02 and a visual magnitude of 1.25, this gives it a B-V colour index of −0.23. Stars with negative B-V magnitudes are blue in colour, those with B-Vs of 1 or 2 are red in colour and those with indices between are white, yellow or orange.

Once you have used a telescope for any length of time you will know the difficulty involved in specifying exactly where in the field of view to look, especially if you have tried to show objects to non-astronomers. Saying 'up near the top' or 'on the left' is too vague, especially when you consider that sometimes the view through the eyepiece will be reversed while at others it will not. To avoid the problem, astronomers use the terms 'leading' and 'trailing' to describe objects. If your telescope is not moving, the stars will drift in one side of the field of view and out the other. The side they are coming in is the eastern side, the trailing edge. The other side is the western or leading side. 'North' and 'south' are also used with south being towards the south celestial pole and north being the opposite direction. When it comes to planets there is a different meaning for east and west. Before the beginning of space exploration, the eastern side of a planet was the side towards east on Earth—the trailing edge; now the eastern side is the same as on Earth—the direction in which the Sun rises, usually the leading edge. Just because the scientists changed the definition doesn't mean that they changed the names of places, though. The Mare Orientalis (The Eastern Sea) is now found on the western side of the Moon.

One of the first tasks given to professional astronomers was to determine an accurate time system for use by the general public—even today our clocks and calendars are kept accurate by very precise astronomical observations. Because astronomers work all around the world, observing the same objects at different times of day, they use a system of time measurement that is the same all around the world. It is known as Universal time, or UT and is recorded in 24 hour format, that is 17:00 instead of 5:00 pm. Universal time is the same as Greenwich Mean Time (GMT), but is called universal to remove any nationalistic connotations. When astronomical observations are made they are always recorded with the universal time and date. Table 6.2 (see following chapter) gives the differences between local and universal time for some locations. It is important to note that the date used in Universal time changes at 0:00h UT, so 2:00am on 6 October 1986 in Christchurch, New Zealand is 14:00 on 5 October 1986 UT

Another problem astronomers face is the difference between time measured by the stars and time measured by the Sun. As we have seen in chapter 1, it takes the Sun 24 hours to appear to travel once around the Earth whereas it takes the stars only 23 hours 56 minutes 4 seconds. Although this is a difference of only 3 minutes 56 seconds each day, it is enough to make a large difference to the accuracy of astronomical observations. To solve this problem astronomers use a system of time that matches the motion of the stars, sidereal time (ST). One sidereal day has 24 hours in it just like a normal day, but these 24 hours are squeezed into 23 hours 56 minutes 4 seconds of normal time. A clock set to run at sidereal time would appear to run about 0.3% fast.

Because times are written so often in astronomy, a shorthand method is used. Instead of writing 11 hours 50 minutes and 7 seconds for example, an astronomer would write 11h50m07s to save time and space. Another point to be careful of is writing dates. In Australia and New Zealand we normally write the date in the order: day, month, year. In the United States the usual form is month, day, year while the Japanese use year, month, day. To avoid confusion it is better to write the name of the month rather than just a number. 4 Jan 1987 is preferable to 4/1/87 which could also mean April Fool's Day in the United States.

There are some astronomical objects, such as variable stars, which are observed on and off over a long period of time; some variable stars have records going back over a hundred years. This creates difficulties when working out the length of time between two observations. For example, how many days are there between 5 March 1986 and 8 October 1986? To remove the need to constantly do complex calculations to work out such details, astronomers use a system of dates known as the Julian system. The Julian date of any day can be fairly easily calculated from tables or by the method given in appendix 1. It is the number of days since noon on 1 January 4713 B.C. This date was chosen as being far enough back in time for there not to be any

Table 5.1 Method for calculating the Julian date

year	Julian date	year	Julian date
1981	2444604	1996*	2450082
1982	2444969	1997	2450448
1983	2445334	1998	2450813
1984*	2445699	1999	2451178
1985	2446065	2000*	2451543
1986	2446430	2001	2451909
1987	2446795	2002	2452274
1988*	2447160	2003	2452639
1989	2447526	2004*	2453004
1990	2447891	2005	2453370
1991	2448256	2006	2453735
1992*	2448621	2007	2454100
1993	2448987	2008*	2454465
1994	2449352	2009	2454831
1995	2449717	2010	2455196

month	normal	leap
January	0	0
February	31	31
March	59	60
April	90	91
May	120	121
June	151	152
July	181	182
August	212	213
September	243	244
October	273	274
November	304	305
December	334	335

Note: To use this table do the following:
1 Find the Julian date for the year
2 Add the number for the month. Remember to use the leap table for years marked *
3 Add the date.
4 Use universal time and date for the calculations. Subtract 1 if the UT is before 12 noon.

ancient astronomical records made before that date. Table 5.1 gives a method of calculating the Julian date for any time in the next few years. Appendix I gives a far more detailed method of calculating this date.

The calendar which we take for granted has had a fairly chequered career. It wasn't until 45 B.C. that any attempt was made to bring in a uniform system throughout the world. At this time Julius Caesar brought in a calendar with leap years: the Julian calendar, to be used throughout the Roman Empire. This calendar used an extra day every four years to approximate the length of time the Earth takes to go around the Sun. Unfortunately, the 365¼ days this gives is far enough out to cause the calendar to run a little fast. By 1582 this discrepancy had caused the calendar and the Earth's motion to differ by ten days. It was in 1582 that Pope Gregory decided that this had to be fixed. He started by removing the leap day from years ending in 00 if their first digits were not divisible by four. For example, 1700, 1800 and 1900 were not leap years, but 2000 will be. This system, which we still use today, is known as the Gregorian calendar. The most controversial step the Pope took was to remove the days 5 to 14 October 1582. This action caused riots throughout Europe as people thought they were being cheated of ten days of their lives. Some countries did not follow the Pope's edict. England didn't change until 1751; Russia didn't make the adjustment until after the Revolution in 1918—by this time the error had grown to thirteen days.

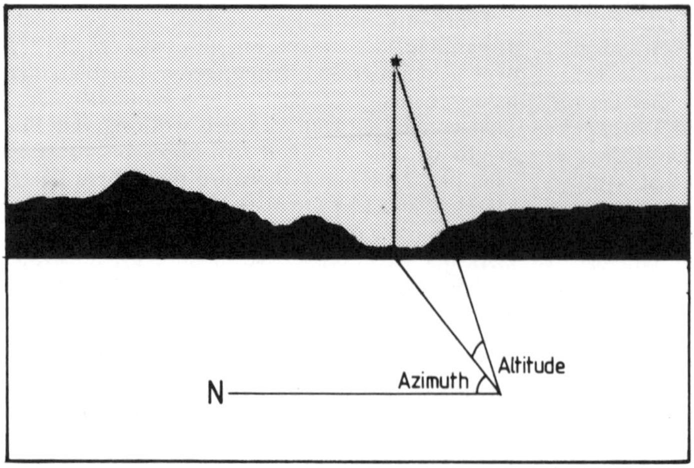

28 This diagram shows the method used to calculate the altitude and azimuth coordinates of a star. Altitude is measured upwards from the horizon and azimuth is measured around, via east, from north. In this case the altitude would be about 20° and the azimuth about 80°.

Another fairly recent change to the calendar was the start of the year. Up until 1622 the year started on March 25—this was for religious reasons, March 25 being exactly nine months before Christmas. Finally, the start of the day was changed. Until quite recently astronomers and mariners started their days at noon instead of midnight. Having the day start in the daytime was preferable for astronomers as it meant that all of a night's observations had the same date. Sailors also preferred this system for the keeping of the ship's log. It wasn't until about 1925 that everyone changed to match the normal civil practice of a midnight start.

Distances often need to be measured and expressed in astronomy, very large distances usually. Using the units of length we use on the Earth means that you very quickly end up with huge and unwieldy numbers, for example, the distance to the nearest star is 41 000 000 000 000 kilometres. The distance to the nearest large galaxy has six more zeros in it. To remove some of the difficulties associated with these large numbers, astronomers use larger units than the kilometre, the most common being the light-year and the parsec.

Oddly enough, the light-year has nothing to do with the measurement of time—it is solely a unit of distance. Light is the fastest thing in the universe: it travels at a constant speed of almost 300 000 kilometres per second in a vacuum. The light-year (written l.y.) is the distance a beam of light would travel in one year. This has been calculated by scientists to be 9 460 350 000 000 kilometres.

The other large unit of distance, the parsec (written pc), was defined by astronomical observation to be 30 857 000 000 000 kilometres. A unit that is sometimes used for measurements within the solar system is the astronomical unit or AU. This is the average distance from the Earth to the Sun and is equal to 149 600 000 kilometres.

Angles are the most commonly measured quantity in astronomy, for it is by angles that the positions of all the objects in the sky are described. The angles used are exactly the same as those used in school geometry and it takes just a little while to learn all about them. The Babylonians were the first to formalise the writing of angles, they divided the circle up into 360 equal parts that we now call degrees, written as °. They divided each degree into 60 parts called minutes, written ′, and then made each minute into 60 smaller segments called seconds, written ″. For example, the angle 45 degrees 15 minutes 47 seconds is written 45°15′47″.

The simplest way to describe the position of an object is to say how high up it is, and what direction it is in. This is called the altazimuth system. The name comes from two words, altitude, the height above the horizon and azimuth, the direction. Altitude is measured from 0° at the horizon to 90° directly overhead. Estimating the altitude of an object is quite simple if you remember one point. An object that appears to be halfway up the sky has an altitude of 30° not 45° as would be expected. This is because we tend to stretch angles close to the horizon and compress angles overhead, it is a psychological effect and has been well studied. There is nothing you can do to change it, just take it into account when you are observing.

Azimuth is easily determined if you know which way north is. There are all sorts of methods for determining which direction north is, from using the stars, as we will see in chapter 6, to using a magnetic compass. I find the easiest method to use in the city is a street directory. Most directories have north marked on them; you can use this to work out north for your location. Once you know which direction north

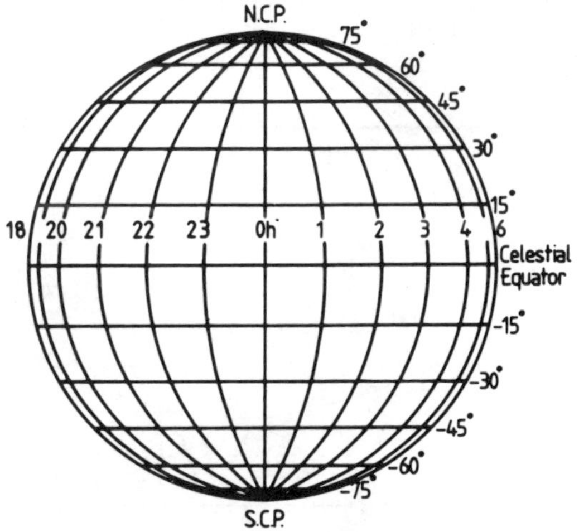

29 A diagram of one half of the celestial sphere showing lines of declination and right ascension. Note the similarity between it and the lines of latitude and longitude on the Earth.

is in you can work out all the other compass points easily. Azimuth is measured in degrees starting with 0° at north and working around through east to south and west and back to north. This means east is 90°, south is 180° and west is 270°. Azimuth is always measured towards the east with 359° being just one degree west of north.

Another use for angles is to describe the size of astronomical objects or their separation from other objects. Because many of these measurements are extremely small most of these angles will be in seconds of arc. Through the telescope, you will quickly get used to these measures, and they will make observing much easier.

To start using these measures, look at the Moon. It is 30' in diameter. When you are using your telescope or binoculars observe object 59, the double star μ Crucis. The two stars in this object are separated by 34.9″. Try using different eyepieces in your telescope and you'll get some idea of the size other objects will appear.

As the Earth spins in space, there are a number of positions that astronomers can define to help them locate the positions of objects. The first are the points above the north and south poles. Because the Earth is spinning about the poles, the points in the sky above them do not appear to move. Astronomers call these two fixed points the north and the south celestial poles, or NCP and SCP for short.

On the Earth the line that is halfway between the two poles is called the equator; it separates the Earth into northern and southern hemispheres. In the sky astronomers have called the line halfway between the celestial poles the celestial equator. It divides the sky into a northern and a southern hemisphere also. On Earth the area between the equator and the poles is further divided by lines of latitude so that the exact position of a town or city can be determined. Similarly, the area of sky between the celestial equator and the celestial poles is divided by lines of declination (Dec.). Like latitude, declination is measured in degrees from the equator. Declinations that are north of the celestial equator are positive, those to the south are negative with the poles being +90° and −90° respectively.

In addition to declination, another quantity is needed to uniquely specify the position of any astronomical object. Lines of right ascension (RA) which run from the NCP to the SCP are used for this. These lines are the counterpart of the lines of longitude that run from the Earth's north pole to its south. Unlike latitude however, there are no obvious starting points for the measurement of longitude. The British used a line running through the Royal Observatory at Greenwich as a starting point, the French used a line through Paris, other nations used lines though Cadiz, Naples and Rio de Janeiro to name a few. For many years all these rival systems were in use, but on 13 October 1884 all

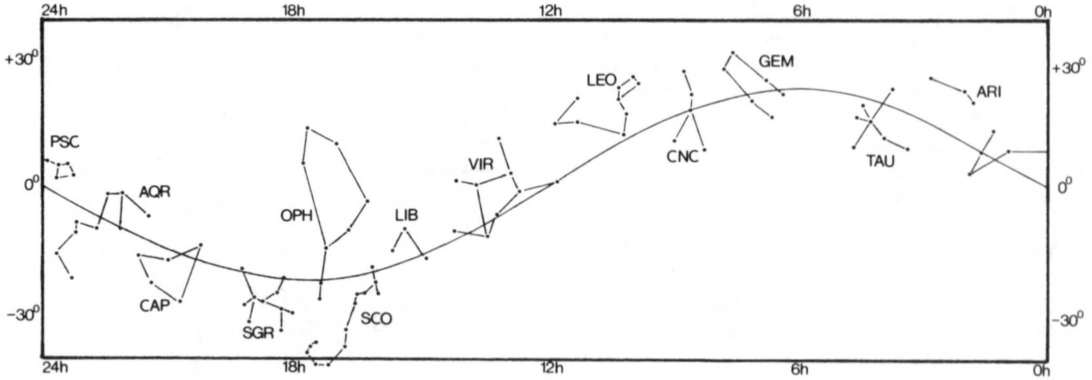

30 This map of the heavens shows the ecliptic. You will notice that the ecliptic passes through thirteen constellations, the twelve signs of the zodiac and Ophiucus. The Sun and planets are always seen in the area of the sky marked by the ecliptic, though Pluto can be found at some distance from the others.

the maritime nations agreed that the system based on Greenwich should be the only one used throughout the world.

Astronomers were faced with a similar situation when it came to the lines of right ascension, but as there were no national arguments to settle, they chose on scientific grounds. Each year, as the seasons change, the Sun moves higher and lower in the sky, twice a year crossing the celestial equator. These crossings occur in March, on the vernal equinox, and in September, the autumnal equinox. Astronomers decided to use the point of the vernal equinox as their starting point. When the Greek scientist Hipparchus introduced this system, the point of the vernal equinox was in the constellation of Aries, so the point was known as the First Point of Aries.

As the sky appears to revolve throughout the night, it can be used as a clock, a very accurate one in fact. Throughout the centuries observations of the stars have been used to keep time pieces accurate. The ancient astronomers also understood this, and so instead of numbering right ascension in degrees minutes and seconds, they numbered it in hours minutes and seconds. In the whole circle of the sky there are twenty-four hours of right ascension, each hour has sixty minutes, each minute sixty seconds just as normal time does. It is important not to confuse the minutes and seconds of declination and the minutes and seconds of right ascension, so they are normally written differently. As an example, the star α Centauri has a right ascension of 14h38m18s and a declination of $-60°45'28''$. Together the system of right ascension and declination is known as the equatorial coordinate system.

The time system of right ascension is used for more than just specifying the position of an object. Local sidereal time is defined to be the value of right ascension that is directly overhead. You can work out if an object will be visible in the sky if you know its right ascension and the sidereal time. If the RA of the object is within a few hours of the local sidereal time, then the object will be above the horizon. Equatorial coordinates have a very big advantage over the altazimuth system. Whereas the altazimuth coordinates of an object will change from moment to moment as it moves across the sky, the equatorial coordinates do not change—therefore objects are usually listed by their equatorial coordinates.

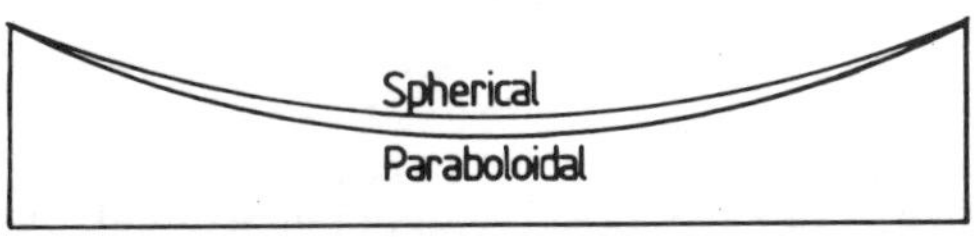

31 The difference in shapes between paraboloidal and spherical mirrors is much less than that shown in this diagram. When making a mirror, opticians aim for a spherical shape and then deepen the centre to form a paraboloidal surface.

If you have ever played with a spinning top, you will know that it leans over and wobbles; this wobbling is known as precession. The Earth's rotation is also affected by precession, it rocks very slowly from side to side, with one full wobble taking 25 800 years. During this period the poles of the Earth trace out large circles in the sky. Because of this the equatorial coordinates of every object slowly change. In the 2100 years since equatorial coordinates were first used, the starting point, the first point of Aries, has moved out of Aries and into Pisces and will soon move on into Aquarius.

Because of this slow shift of positions, the coordinates of an object are always specified for a given time or epoch. The most common epoch used in astronomy is 1950.0, that is the positions that the stars were in on 1 January 1950. This is known as the standard epoch. There has been some move to change the standard epoch to 2000.0, but there has been no official word on this so far. In some astronomical books you will find 'epoch of date' used. That means that the positions given have been corrected for precession up until the date given for the observations. Sometimes the positions for comets are given in epoch of date, but normally they are given in 1950.0 coordinates as well.

The Astronomical Almanac uses the epoch of the year it is written for, and gives us a good idea of the speed of precession. For example, the position for Sirius (α Canis Majoris) in epoch 1967.0 is 6h43m41s $-16°40'10''$, by epoch 1977.0 this has changed to 6h44m08s $-16°41'01''$. This isn't a large amount, it is certainly much too small to be seen with the naked eye, yet it is enough to be easily measured with professional equipment.

Equatorial and altazimuth coordinates are not the only systems of specifying the position of an object, but they are the most common by far.

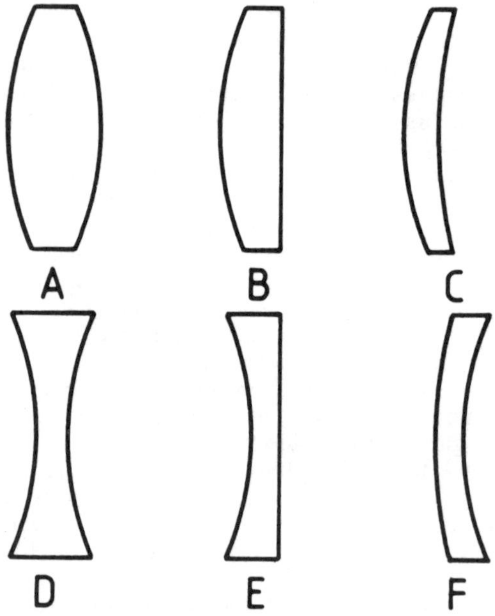

32 Some of the lenses commonly used in astronomical instruments and other optical systems. A: Biconvex, B: Plano-convex, C: Concavo-convex, D: Biconcave, E: Plano-concave, F: Convexo-concave.

One other system that is used by professionals is the galactic coordinate system. This again works very much like latitude and longitude, but instead of using the equator and the poles of the Earth, it uses the poles of the galaxy and the galactic plane as its reference points. Objects are then given a galactic latitude and galactic longitude. It would be very rare to come across this form of position in any amateur astronomical book, but you may find it in professional publications.

Precession is not the only effect that makes the stars' positions change over time. Although they do it very slowly, the stars do move around in the sky. This motion is called proper motion and although the stars appear fixed to the naked eye for thousands of years, it is taken into account by astronomers when they are calculating accurate positions. The star with the greatest proper motion is a faint red star in the constellation of Ophiucus. Barnard's Star has an annual motion of 10.29″ of arc, that is 1° in just 350 years. If you want to see Barnard's Star for yourself it has been included in our selection of objects.

Another important line that astronomers have drawn in the sky is called the ecliptic. This is the line that the Sun follows as it moves through the sky. Because the solar system is fairly flat, all the planets, except Pluto, also appear in the sky fairly close to the ecliptic. This line passes through thirteen constellations as it goes around the sky. Twelve of these constellations are the zodiac—the thirteenth, Ophiucus, is not included in the zodiac, probably because astrologers don't like the number thirteen. When people talk about star signs, what they are referring to is the constellation that the Sun was in when you were born. The problem with this is that the zodiacal constellations are not spread throughout the sky as evenly as astrologers spread the dates throughout the year. Another difficulty is that precession has moved the constellations' positions over the thousands of years since the system was invented, so the places now used are no longer accurate, they're one or two constellations out.

Like astronomers, opticians have also developed their own ways of describing optical instruments. Lenses and mirrors in particular have special descriptions. A surface that curves inwards as a spoon does when held for eating, is called a concave surface. A surface that curves outwards, like an upside down spoon, is called convex. Surfaces that have more complicated curvatures are called complex. Lenses and mirrors are described by the shape, or shapes, of their surfaces. A biconvex lens has two convex faces, a plano-concave lens has one flat surface and one concave surface, and so forth.

Because nearly all telescope mirrors are concave they are often named after the exact shape of their surface. There are a very important set of mathematical shapes that derive from a cone; most telescope mirrors have one of these shapes. The mathematical shapes are the circle, the ellipse, the parabola and the hyperbola—they are named depending upon the exact way they curve. When they are applied to telescope mirrors they are rotated about the mirror's axis and renamed. The circle becomes spherical; the ellipse, ellipsoidal; the parabola, paraboloidal; and the hyperbola, hyperboloidal. Figure 31 shows the differences in some of these shapes.

There are a number of other terms in astronomy which have simple meanings; most of these are included in the glossary at the back of this book. If you come across a new word, try looking it up there. At first glance there seems to be an impossible number of concepts and terms to be learnt; it is not necessary to learn them all at once. As long as you know that the terms exist you will become familiar with them as you do more astronomy.

6

Getting ready to observe

Once you have purchased a telescope, it is only natural that you will want to use it at the first available opportunity—it is now that you will first encounter Murphy's Law as it applies to astronomy. It is always cloudy on the night you buy a new telescope. Even if it isn't cloudy, there are a few checks and adjustments you will need to make before you can use the telescope properly.

Firstly, the telescope must be assembled. Most telescopes that require assembly come with a good set of instructions, usually written in Japanese-English. Follow the instructions carefully and if you have any trouble contact the store from which you purchased the telescope. This is one advantage of going to a specialist shop: they usually know something about the items they are selling. Refracting telescopes and Cassegrains are fairly easy to set up, but Newtonians sometimes need to be collimated.

Collimating a telescope means adjusting the position of the optical elements so that they are all on the same optical axis. This can be done fairly simply for Newtonians, but refractors and Cassegrains need to be done by an optician. To collimate your telescope, do the following: remove any eyepiece from the drawtube and, holding your eye back a little from the end of the drawtube, look in. The first thing you should see is the small secondary mirror. In the middle of this you should see the reflection of the large primary mirror, this in turn will have a reflection of the secondary mirror, and in the centre of the lot will be a reflection of your eye. If all these reflections are concentric, then the telescope is collimated and you need not worry.

If the reflections are not concentric, look at figure 33 to determine whether it is the primary or secondary mirror that needs adjusting. The primary mirror of a telescope is moved by very gently turning the screws underneath the mounting cell. The secondary mirror is adjusted either by turning the mounting screws on the back of the holder, or by adjusting the screws that hold the secondary cell in the tube. When making adjustments do them carefully, and only a little at a time—although it may look like the mirrors are a long way out, the adjustments needed are usually minute. When you are collimating continually check your work, as it is easy to go too far and then have to correct your mistake.

The other part of the telescope that will need to be adjusted before you can do any observing is the finder scope. Not only will it need to be lined up when you first buy the telescope, but it will also need aligning after the telescope has been stored or transported as they are notorious for getting knocked and bumped. Aligning a finder scope is best done during the daytime, but be very careful to avoid pointing the telescope directly at the Sun. Many finders have a system of locking nuts to help keep them aligned; if these are fitted make sure that you release them completely before trying to do any adjustments, otherwise you may strip the thread on the mounting screws.

Choose a small object at least a kilometre away from you, a television antenna, or a chimney or some other feature of a distant building is suitable. *Do not* use the Sun. Put a low powered eyepiece in the telescope, then looking through

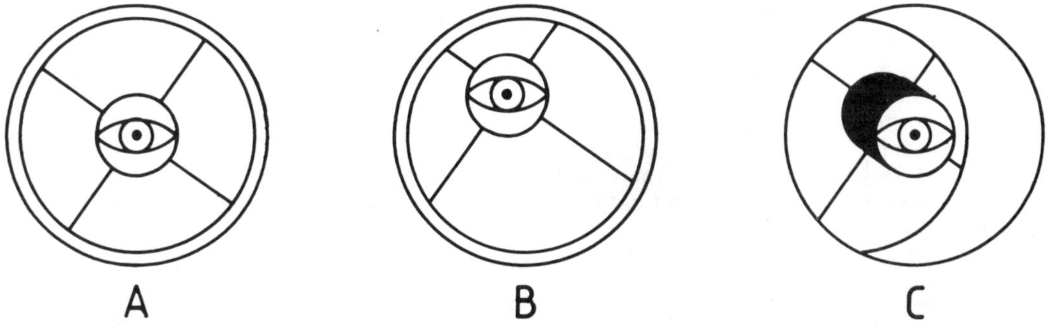

33 When collimating a Newtonian reflecting telescope it is important to ensure that all the optical elements are on the same optical axis. By comparing the view through the empty drawtube of your telescope with the three diagrams above you can correctly adjust your scope. Diagram A shows the correctly adjusted view. Diagram B indicates that the primary mirror needs adjusting. Diagram C indicates that the secondary mirror is out of alignment.

the telescope, point it at your chosen object. Carefully remove the eyepiece and replace it with a high powered one. Adjust the position of the telescope so that the object is in the middle of the field of view. Now, without moving the telescope, look through the finder scope. The distant object should be centred on the cross-hairs. If it is not slowly adjust the mounting screws on the finder so that the object is centred. Tighten the locking nuts if they are attached. When you are doing the adjustment, occasionally check that the main telescope is still pointed at the object and that you haven't accidentally bumped it away.

With this adjustment made you are almost ready for your first night's observing. If you are unfamiliar with using a telescope, and you have one with an equatorial mount, I would suggest that you don't try setting the mount up correctly to begin with. Setting an equatorial mount is difficult and there is no reason why you should do it on your first night. Just use the mount to support the telescope so that you can do some viewing.

After a few sessions observing and getting used to the telescope, it is time to set the equatorial mount correctly. This is a complex task the first time you do it, so give yourself plenty of time. It is better to begin the operation in the afternoon before you observe, when there is plenty of light about. If you will only use the telescope for viewing then you need not be very accurate, but if you want to do photography then you must make each adjustment very carefully to ensure that the mount is aligned as well as possible.

Before you begin you will need a few things; firstly an accurate north-south line. Appendix 2 gives a simple method of constructing such a line using the Sun. You will also need a spirit level and a specially cut triangle of cardboard. You can make the triangle quite simply from any stiff cardboard. Find out your latitude—table 6.1 gives this for some cities—and draw an angle on the cardboard equal to this. Mark the angle as accurately as possible using a protractor. Next make a third line forming a right-angle triangle as shown in figure 34. Carefully cut the triangle out and mark the angle which is equal to your latitude.

Take your equatorial mount and roughly align it so the polar axis points south. There is usually a locking nut which holds the head of the mount in position; loosen this nut so that you can more accurately align the telescope north-south. The more accurately you do this adjustment, the easier the telescope will be to use when you are observing. Another adjustment that you must make on the telescope head is its tilt. There is usually some device which keeps the tilt locked once you set it, remember to release this mechanism before you try to adjust the head. The tilt of the polar axis must be set equal to your latitude. Take your piece of cardboard and place it upright on the telescope's polar axis with the angle of latitude at the top; this is shown in figure 35. Using the spirit level adjust the tilt of the head so that the top of the cardboard is level. Retighten the locking mechanism to keep this setting; if you always observe from the same spot you will not need to make this adjustment again.

Because of the way it is arranged, the German equatorial mount uses a counter-weight to balance the telescope; you will need to set it also before you can use the telescope. With the telescope in position, turn it about the polar axis so that the declination axis is horizontal. Now, with all the locking devices loosened, move the counter-weights up and down the shaft until the telescope balances perfectly. Make sure that you have an eyepiece in the drawtube, and that you have removed all the lens caps and covers. I remember once spending fifteen minutes balancing my telescope only to find that I'd left the metal mirror cover in place and had to start all over again. Once you have balanced the telescope it should not need to be done again unless you attach a camera or other heavy accessory to the scope.

With the polar axis aligned and the telescope balanced, you have done as much as you can until night time. The final adjustment to be made is to set the circles on the telescope so that they can be used to find objects. Point the telescope at a bright star and centre it in the eyepiece. Check the list in table 6.2 to find the star's right ascension and declination. Loosen the locking nut and turn the declination setting circle so that it reads the declination of the star. If you have a motorised mount, and it drives the right ascension circle, simply set the circle to read the RA of the star.

If your mount is not motorised, or if it doesn't drive the right ascension setting circle, you will need to set the circle to read the hour angle of the star rather than its right ascension. To calculate the hour angle of a star you need to know its RA and the local sidereal time. Subtract the sidereal time from the star's RA—the number you get is the hour angle. If the number is negative, add twenty-four hours to it. Set the right ascension setting circle to read the hour angle. The mount is now fully set for observing.

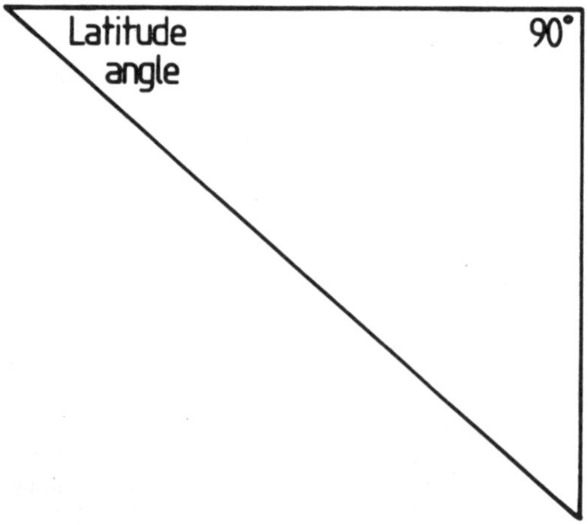

34 A triangle suitable for use in setting up an equatorial mount can be drawn quite simply as shown above. This one has been drawn for Wellington, New Zealand. When constructing your own version of the triangle it is important to be as accurate as possible in setting the latitude angle.

At night it is a simple task to find south using the stars. The method I will explain is accurate enough to roughly align a telescope for viewing, but not good enough for photography. The first thing you will need to do is find the constellation of the Southern Cross. This constellation is visible all year round from much of the southern hemisphere. If you are having difficulty, look for two bright stars called the pointers. These are α and β Centauri and they point towards the bright star at the top of the cross. Once you have found the cross, identify the stars in it according to figure 37. Then draw a line from the top star to the bottom star. Continue the line for four and a

Table 6.1 Latitude and longitude positions for some important towns (southern hemisphere)

Location	Latitude	Longitude	Time Zone
Australia and the Pacific			
Adelaide, Australia	34°56′S	138°36′E	+9.5h
Alice Springs, Australia	23°42′S	133°52′E	+9.5h
Auckland, New Zealand	36°55′S	174°47′E	+12h
Brisbane, Australia	27°30′S	153°00′E	+10h
Cairns, Australia	16°51′S	145° 43′E	+10h
Canberra, Australia	35°18′S	149°08′E	+10h
Christchurch, New Zealand	43°33′S	172°40′E	+12h
Darwin, Australia	12°23′S	130°44′E	+9.5h
Dunedin, New Zealand	45°52′S	170°30′E	+12h

Table 6.1 continued

Location	Latitude	Longitude	Time Zone
Hobart, Australia	42°54′S	147°18′E	+10h
Honiara, Solomon Islands	9°28′S	159°57′E	+11h
Kaitaia, New Zealand	35°08′S	173°18′E	+12h
Kingston, Norfolk Island	29°04′S	167°57′E	+11.5h
Melbourne, Australia	37°45′S	144°58′E	+10h
Noumea, Nouvelle Caledonie	22°16′S	166°26′E	+11h
Nuku'alofa, Tonga	21°09′S	175°14′E	+12h
Papeete, Tahiti	17°32′S	149°34′W	−10.5h
Perth, Australia	31°58′S	115°49′E	+8h
Port Moresby, Niugini	9°30′S	147°07′E	+10h
Port Vila, Vanuatu	17°45′S	168°18′E	+11h
Suva, Fiji	18°08′S	178°25′E	+12h
Sydney, Australia	33°55′S	151°10′E	+10h
Wellington, New Zealand	41°17′S	174°47′E	+12h
Africa			
Antananarivo, Madagascar	18°55′S	47°31′E	+3
Brazzaville, Congo	4°16′S	15°17′E	+1
Cape Town, South Africa	33°55′S	18°22′E	+2
Dar es Salaam, Tanzania	6°48′S	39°17′E	+3
Gaborone, Botswana	24°45′S	25°55′E	+2
Harare, Zimbabwe	17°50′S	31°03′E	+2
Johannesburg, South Africa	26°13′S	28°02′E	+2
Kinshasa, Zaire	4°18′S	15°18′E	+2
Luanda, Angola	8°48′S	13°14′E	+1
Lusaka, Zambia	15°25′S	28°17′E	+2
Maputo, Mozambique	25°58′S	32°35′E	+2
Nairobi, Kenya	1°17′S	36°49′E	+3
Windhoek, Namibia	22°34′S	17°06′E	+2
South America			
Asuncion, Paraguay	25°16′S	57°40′W	−4
Brasilia, Brazil	15°47′S	47°55′W	−3
Buenos Aires, Argentina	34°36′S	58°27′W	−3
La Paz, Bolivia	16°30′S	68°09′W	−4
Lima, Peru	12°03′S	77°03′W	−5
Montevideo, Uruguay	34°53′S	56°11′W	−2
Port Stanley, Falkland Is	51°42′S	57°51′W	−4
Porto Alegre, Brazil	30°04′S	51°11′W	−3
Rio de Janeiro, Brazil	22°54′S	43°14′W	−3
Santiago, Chile	33°27′S	70°40′W	−4

The prime meridians for the time zones given are as follows

0	0°W		
−1	15°W	+1	15°E
−2	30°W	+2	30°E
−3	45°W	+3	45°E
−4	60°W	+4	60°E
−5	75°W	+5	75°E
−6	90°W	+6	90°E
−7	105°W	+7	105°E
−8	120°W	+8	120°E
−9	135°W	+9	135°E
−10	150°W	+10	150°E
−11	165°W	+11	165°E
−12	180°W	+12	180°E

The prime meridians for zones with half hour differences are those for the full hour before them on the list, plus 7°30′.

Note: This is by no means a complete list, but is enough to cover most of the population of the region. If none of the places listed is near to your location,a good atlas will probably give you a reasonable idea of your exact coordinates. The times given are hours ahead (+) or behind (−) Universal Time. No account has been made for daylight savings—this usually results in an extra hour being added to the time given.

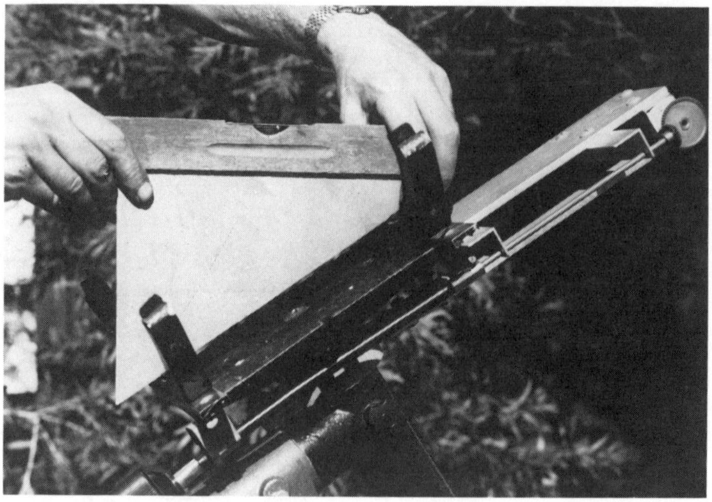

35 Using a triangle to set up an equatorial mount. The triangle is held in contact with the bed of the mount and a small spirit level is held on top. By using the latitude adjustment the mount can be set to the correct inclination. PHOTO: DAVID REIDY

half times the distance between the two stars. This point is very close to the SCP.

Use the azimuth adjustment on the mount to point the head in the direction of the SCP. Now make sure that the telescope is sitting on its mount so that it is directly above the counter-weight and the declination reads 90°. While viewing through the finder, turn the equatorial head around in the azimuth until the region you identified is centred. Lock the adjustment and you are ready to observe.

The first problem you will encounter in using the telescope is getting it to point in the correct direction—especially if you have an equatorial mount. It is worth spending a little time just getting used to the way the telescope moves on its mount. Once you have a feel for its movements you can try pointing the telescope at an object. Remember, *never* point a telescope directly at the Sun. A cheap magnifying glass can set fire to a piece of paper very quickly, an accurate astronomical instrument can do it even quicker. Through a telescope the Sun's rays will instantly destroy part of the retina of your eye. This damage is irreversible and you would end up partially or totally blind.

If you have a straight-through finder, it is very simple to point the telescope the way you want to. Find the object in the sky if it is bright enough, or some feature that is close to it. Then, keeping both eyes open, look through the finder. Move the telescope so that the views with both eyes line up, that is, the star seen through the finder and that seen without appear as one. Once this has been done the star should be in the field of view of the telescope, provided that the finder scope has been properly aligned.

Before you can use the telescope for viewing anything, it will have to be focused. Focusing means that you will have the best and clearest view possible with the telescope. To focus the telescope you will need to point it towards some object. With a low power eyepiece, any region of

Table 6.2 Bright stars: positions for epochs 1950.0 and 1988.0

Star		RA	Dec.	RA	Dec.
		epoch 1950.0		epoch 1988.0	
α	Eri	1h35m	−57°29′	1h37m	−57°18′
β	Ori	5h12m	−8°15′	5h14m	−8°13′
α	CMa	6h43m	−16°39′	6h45m	−16°42′
α	Leo	10h05m	+12°14′	10h08m	+12°42′
α	Cru	12h24m	−62°49′	12h26m	−63°02′
α	Cen	14h36m	−60°38′	14h39m	−60°43′
α	Sco	16h26m	−26°19′	16h29m	−26°24′
λ	Sco	17h30m	−37°04′	17h33m	−37°06′
α	Aql	19h48m	+8°44′	19h50m	+8°50′
α	PsA	22h55m	−29°53′	22h57m	−29°41′

Note: To use this table for adjusting the setting circles on your telescope, select the bright star from the list, point the telescope towards it and adjust the setting circles to the values given. Use epoch 1950.0 if you are working with positions given for that epoch, otherwise use the epoch 1988.0 figures for epoch of date.

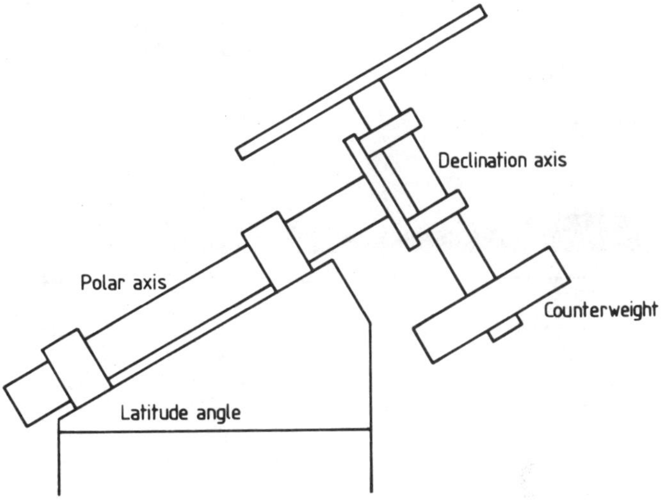

36 The major parts of a German equatorial mount.

the sky will provide enough stars to use. With the eyepiece in position and the telescope pointed skywards, look in the eyepiece, keeping your eyes relaxed. Carefully adjust the focusing control until the stars appear as small points of light. You will need to make a few large turns of the control to get the view nearly correct, then some small adjustments to get it just right. When you are doing this you will need to let the telescope stop shaking between adjustments.

It is possible to focus the telescope so that you can see the stars as disks and all sorts of features moving around on their surfaces. Interesting as the sight may be, it is an illusion. The disk is simply caused by the telescope being out of focus. The features moving about it are caused by small particles of dust in your eye and your eyelashes catching some of the light, not huge 'star spots' or other stellar phenomena.

When you are looking for an object it is best to use the lowest powered eyepiece you have. Not only will this give the brightest view, it will also show you the greatest amount of sky. Using low power, centre the object in the field and carefully change eyepieces to the power you want to use to observe. Remember that the highest power is not always the best to observe with.

Once you have your telescope pointed at the object you will need to constantly adjust its position so that the object remains in the field of view. The higher the power you use the more often you will need to make these adjustments. On an altazimuth mount, or a non-aligned equatorial mount you will need to adjust both axes of the telescope on an aligned equatorial you will only need to move the polar axis. A motorised mount will of course take care of the polar axis adjustments for you, but you may still occasionally have to make manual adjustments as well.

Selecting a site to observe from is more difficult if you are using a telescope, as there are more points to consider. In addition to being away from any interfering lights, the site at which you put a telescope must be sheltered from the wind. Even a light breeze will cause some telescopes to move on their mounts, so unless you have a very sturdy mounting for your telescope, try to avoid windy areas. If your telescope has a motor drive then you will need power to use it. If you cannot arrange mains power for the drive you will have to buy an inverter. This is a device which can turn the twelve volts DC from a car battery into two hundred and forty volts AC for the motor.

A site close to home is the best, but unfortunately there is usually some problem with lights. Try to find somewhere with a minimum of uncontrolled lighting, certainly away from roads and streetlights. If there are only one or two lights try erecting shields to block your view of them. If the problem comes from a neighbour's house, invite them along to observe one night. After you've shown them a few things, mention how much better it would be without their outside lights. They will usually cooperate.

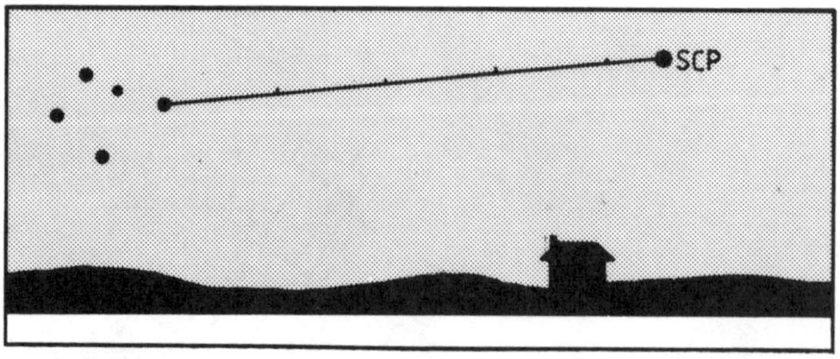

37 Using the Southern Cross it is a simple matter to get a rough position for the south celestial pole. Take a line joining α and γ Crucis and extend it for four and a half times its length. The position you end up with is quite close to the SCP.

Once you begin serious viewing you will want to find better observing sites. The best idea is to contact a local astronomical society; they usually have some favourite observing spots, or perhaps even an observatory of their own. Most clubs are quite willing to help new members or visitors get started.

There are a few simple tips which will make observing much simpler and more enjoyable if you take the time to learn them. The first is the proper way to look through a telescope. When you close one of your eyes, often you will screw the other one up a little as well. Looking into the eyepiece with a partially closed eye can lead to eye strain and poor viewing. The best method of looking through a telescope is to keep both eyes open, one at the eyepiece and one looking at nothing. This keeps both your eyes relaxed and will give much better views. What you see with the eye not at the eyepiece shouldn't bother you. If you've a good observing site, there shouldn't be enough light to see anything in any case. Even if there is enough stray light to see by, it is an easy task to simply direct your attention to only one eye.

Another thing you must avoid is touching the telescope while you are looking. Apart from focusing the eyepiece, there is no need to touch the telescope while you are observing. The mount on the telescope should be strong enough to hold the telescope, touching it will only introduce unwanted movement and vibration to the view. When adjusting the position of the telescope while you are viewing, use only the extension cables. It is a bad habit to get into to move the telescope by pushing the tube as this will loosen the mountings and may strip the threads on the gears if you are not careful.

A problem that often occurs in the evening is dewing. This is the formation of a coating of moisture on any exposed surface. There are a number of ways of reducing the problem if it becomes annoying. Most refractors have a tube which extends out from the objective lens for a distance; this is known as a dew cap. Its function is to prevent the formation of dew on the objective lens. It does this by keeping a relatively still blanket of air above the objective. The most effective length for a dew cap is about three times the diameter of the objective, but most telescopes have one shorter than this. If dewing is a problem with your refractor, you can extend the tube using a cylinder of cardboard, but make sure that it doesn't obstruct the view.

Newtonian reflectors are the least prone telescopes to dewing as their main mirror is safely at the end of a long tube. Sometimes, however, the small diagonal mirror will dew up. If this happens try holding your hand near to it, but not touching. The heat from your hand will often solve the problem. It is important that you don't touch the mirror's surface, as skin prints are impossible to remove from a front coated mirror.

Reflecting telescopes like the Cassegrains are perhaps the worst offenders when it comes to dewing up. Rarely will a manufacturer provide a dew cap for the scope. Besides, an effective dew cap would need to be 300mm long for a 100mm aperture telescope. To remove the need for such devices, amateurs use two techniques. The first involves making a small dew cap and fitting it with a heating element run from a battery. Heating the air above the telescope's corrector plate by only 0.5° to 1.0° will stop dew from forming quite effectively. Heated dew caps can be constructed quite simply; your local society can show you how to do this.

A simpler method than constructing a heating element is to use a small hair drier. Although these usually plug into the mains power it is much more convenient and safer to buy a small rechargeable unit. When dewing occurs you

38 A home-made heated dew cap for use on the 125mm Schmidt camera used to take many of the photos in this book. PHOTO: DAVID REIDY

simply play a stream of warm air from the drier over the surface. Be careful not to heat the corrector too much for this will adversely affect the performance of the telescope. All the parts of the instrument should be at about the same temperature if it is to work well.

There are many different ways of making astronomical observations; we shall look at them in the coming chapters as they relate to different astronomical objects. The requirement to keep good records is necessary for all sorts of astronomical observations, whether they be serious or not. The casual observation you make one night may turn out to be very important if records are needed of an astronomical event. If for example you were to look at a galaxy on one night and notice nothing unusual, and a supernova was discovered in it on the next night, your observation of nothing unusual would help to calculate the exact time of the event.

Even if your records are never used for professional astronomy, they will give you an idea of how you are progressing in the hobby. They may even be valuable at a later time if you decide to join an observing program of some sort. One way of recording observations is to keep them in an exercise book or journal, but a better way is to use a loose leaf folder and individual sheets. The advantage of the loose sheets is that you can take only the number you need with you when you observe. Take a clipboard or something to write on and enough clips to hold the paper at all four corners: having it flap around when you are trying to write is most irritating. You can design an observing record sheet for your own use and photocopy them off so that you'll have personalised records.

Regardless of the method you use to record them, all observations must include the universal time and date of the observation, the telescope being used, the magnification used and the name of the object. The record should also contain a diagram or description of what is being observed. Again the problem arises of providing illumination for reading maps and recording observations. With a telescope the same small red torch I described in chapter 1 can be used. For added convenience a small holder for the torch attached to either the mount or the telescope itself will make it much easier to see what you are doing. Such a holder can be made from a large bulldog clip, or large wooden peg. Make sure that when the torch is in the holder it can't be seen with your eye at the eyepiece. If it can be seen, it will distract you while you are viewing.

A good method of illuminating your records is to use one of the small clip-on lights which are available. These come with a small battery pack which will fit into your pocket. Having the clip means that they can be easily attached to your books or clipboard, and it will also help hold the papers in position. You will need to cover the top part with red cellophane as you did for your torch to lower the level of illumination.

Another factor which needs to be recorded with your observations is the seeing conditions. Seeing conditions describe the state of the atmosphere when you are observing. Air currents in the upper atmosphere act as lenses in bending the light of a star or planet. This causes the image to appear to move. This is what causes the twinkling of stars. There is a saying that you can identify a planet because it doesn't twinkle—this is not true as planets, especially the bright ones, will twinkle just as much as a star.

When you are observing, the motions of the

Object ______________________ Position __________

Date ____________ Time (UT) ____________________

Telescope ________ Eyepiece _______ Mag. _______

Seeing ____

Observations:

39 A suggested header for astronomical observing records. Sheets of paper with this type of heading photocopied on them are most suitable as observing forms.

40 Author David Reidy showing the use of a clipboard and clip-on light to make observational drawings. Note that this set-up with the light not covered with cellophane was used to make the lunar observation shown in figure 54. PHOTO: KEN WALLACE

stars are more pronounced. On a good night the stars will appear fixed and steady; on a very bad night they will dance all over the field of view making most observing impossible. Of course, the greater the magnification you use, the greater the effect is. It is this effect which limits the actual magnification you can use on your telescope to well below the theoretical limit on most nights. If while you are observing, turbulence is causing a problem, just reduce your magnification. If doing this doesn't remove the problem, or it means that the magnification is too low to observe as you would like, there is not much left to to but give up observing for the night.

On an observing record you should write a description of the seeing conditions, or use one of the accepted scales. The most popular scale used by amateurs has five steps: perfect, excellent, good, poor and very poor. Perfect conditions mean that every star image is rock steady, very poor means that it is almost impossible to observe. The meanings of those in the middle are a little vague, but with experience you will have no trouble determining the correct description.

The exact techniques you will use when observing a given type of object are quite varied, and there are many skills to be acquired along the way. In the next few chapters we will look at the different types of objects in the heavens and some of the methods used to observe them. Don't be in too much of a hurry to learn all the ins and outs—it takes time. A lot of the fun of astronomy lies in learning new things and seeing new objects. Just plain sightseeing is not only fun, but by doing it you will learn how to observe without really trying.

7

The planets

There are many different astronomical objects to be seen, but the best place to start is probably close to home, the solar system. Since ancient times astronomers have known of five 'stars' that moved about the sky while the others remained fixed. The Greeks called these πλανητησ (planetes), a word that means wanderers; we call these objects planets.

Other than the Earth upon which we live, there are five planets that can be seen with the naked eye. Two of these, Mercury and Venus, are closer to the Sun than we are. The remaining three are further out than Earth; they are Mars, Jupiter and Saturn. Because they are faint, three more planets had to wait for the invention of the telescope before they were found.

Uranus was discovered in 1781 by a then amateur astronomer, William Herschel, using a home-made telescope. Herschel thought he had discovered a comet, and actually entitled his report 'An Account of a Comet', but when astronomers calculated this new object's orbit, it turned out to be a new planet, the first detected since antiquity. This discovery took the astronomical world by storm as no-one had even suspected that there could be another member of the Sun's family. For some time there was great debate about the naming of the planet; at one time it was nearly named after the king, but good sense prevailed and it was named Uranus, after the father of Saturn in Greek mythology.

Once Uranus had been observed for a reasonable length of time, it was noticed that it did not behave in exactly the way predicted by theory—its orbit was irregular. Calculations by some of the best mathematicians in the world showed that these discrepancies may have been caused by another undiscovered planet. Two mathematicians, J.C. Adams in England and U.J.J. Le Verrier in France independently calculated expected positions for this new planet. In 1846 Adams sent his predictions to the Astronomer Royal George Airy, Le Verrier sent his to J. Galle in Berlin. Airy shelved the report, Galle started observing at once, and within a few nights had discovered the planet. The new planet was named Neptune after the Greek god of the sea and in keeping with the mythological flavour of planetary names.

There were still problems with the movements of the outer planets even taking Neptune into account, so a search started almost immediately for a ninth planet. Percival Lowell spent many years in a fruitless search for the elusive object. It wasn't until 1930 that Clyde Tombaugh, working with Lowell's old telescope, caught sight of this last planet. The planet was named Pluto after the god of the underworld, though some say that the first two letters are to commemorate the effort of Percy Lowell.

All the planets, except for Pluto, always appear in the strip of sky either side of the ecliptic. Pluto, because of its highly tilted orbit, can appear a fair distance from the ecliptic. If you can find the ecliptic and the constellations of the zodiac in the sky then you will have little difficulty finding the planets. Astronomical publications such as *The Astronomical Almanac* or local astronomical society ephemerides and many of the popular astronomy magazines give the

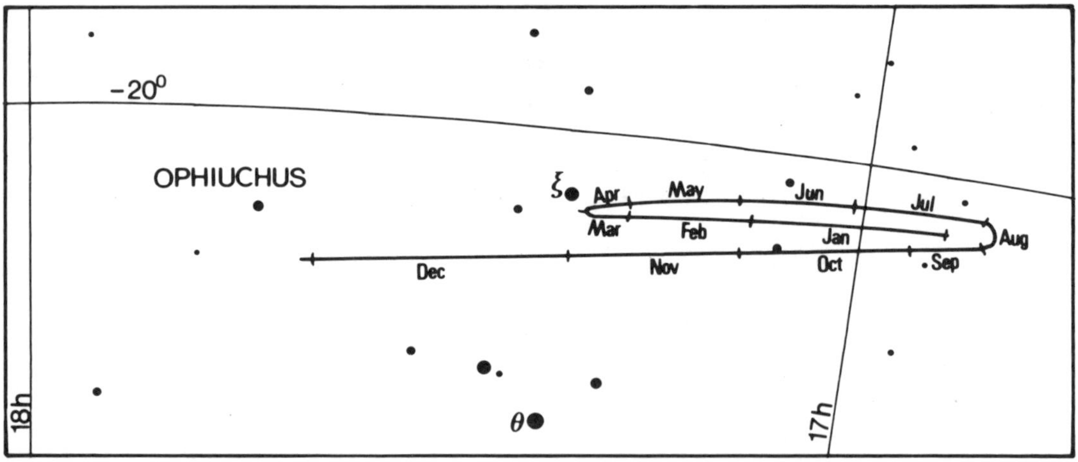

41 This map of Saturn's motion through 1987 is typical of the type of information presented in ephemerides. In this map you can see Saturn's retrograde motion, where it appears to go backwards through the heavens from March to August.

positions of the planets in the sky during the year. If you have a map of where the planets are, it is simply a matter of finding them in the sky. If you only have the positions, then you will need to use your star atlas. Using the right ascension and declination given, plot the position on your map then use that to find it. Chapter 9 explains a method for doing this.

Mercury is by far the hardest of the naked-eye planets to see. It is always close to the Sun in the sky, never moving more than 28° away from it. As astronomical twilight doesn't finish until the Sun is 18° below the horizon, it means that Mercury is rarely seen in a dark sky. The period at which Mercury is at its best lasts only a few weeks, so it is necessary to keep informed if you want to catch it. Because of its distance and its small size, the largest Mercury ever appears is only 9″. You will need to use a fairly high power, 60x at least to make Mercury out as a disk, but the turbulence of the atmosphere near the horizon will make this a difficult task. At maximum brightness Mercury reaches magnitude −1.8, but in the light sky near the Sun it doesn't appear to be very bright.

With a small telescope, there is nothing to see on the surface of Mercury. Seen from spacecraft the surface is very similar to that of the Moon, heavily cratered, but without the large dark areas of the Moon. Like the Moon Mercury shows phases as it moves around the Sun. When it is seen at its best, Mercury looks like a first or last quarter Moon. The only reason for looking for Mercury is to say that you've seen it. It took me a number of years of trying until I was sure that I had seen it properly.

The next planet out from the Sun is Venus; it is much easier to find. After the Sun and Moon, Venus is the brightest object in the sky, reaching magnitude −4.4. Venus is popularly known as the morning or evening star, as it is the first object to appear each night or the last to disappear each morning. In fact, once you know where to look for it you can see it clearly in the middle of the day. Apart from being known as a star, Venus is very popular as a UFO. When it is near maximum brightness and close to the horizon in the evening, air turbulence can make it appear to move in the sky. It doesn't move much, but its little shakes plus a bit of imagination, add up to give some people a UFO. In fact, if you know the time of year when Venus will be in the right position you can start looking for the 'little green men' stories in the afternoon newspapers.

The reason for this amazing brightness is a combination of two factors. The first is that Venus is the closest planet to the Earth: it can come within 41 million kilometres of us. The second factor is the amount of light it reflects from the Sun. Because none of the planets emits its own light we see them only because they

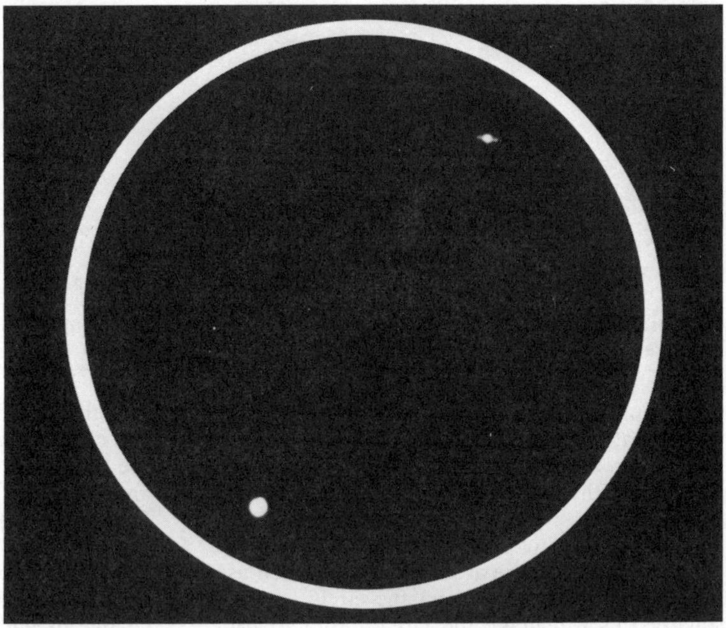

42 Two planets are visible in this photograph, Venus at lower left and Saturn at upper right. If you look carefully you can see the rings around Saturn. The photograph was taken using a 200mm telescope and a 5 second exposure. PHOTO: KEN WALLACE

reflect light from the Sun. The amount of light reflected is called the albedo. Venus has an albedo of 0.76, that means it reflects back 76% of the sunlight reaching it. As a comparison, the Moon reflects only 7% of the sunlight reaching it back towards the Earth.

Venus has such a high albedo because its surface is continuously covered by clouds; in all the years that Venus has been observed a break in the clouds has never been seen. The fluffy white clouds of Venus are totally unlike those of Earth; ours are made of water vapour, Venus has clouds of sulfuric acid. The clouds of Venus hide the hottest planet in the solar system. Although Mercury is closer to the Sun, Venus has a system of trapping heat known as the greenhouse effect.

You will know that if you leave your car parked in the Sun for any length of time with the windows wound up, when you return to it the temperature inside can be very high, up to 60°C. The sunlight coming in through the windows hits the interior fittings and heats them up. As they get hot, they give off heat, but this heat can't escape from the car, so it builds up. This is the same as what is happening on Venus; the arriving sunlight warms the planet. This heat is then trapped by the clouds and so it goes on, reaching temperatures of 485°C, hot enough to melt lead. Mercury on the other hand doesn't have an atmosphere and so no greenhouse effect; its temperature reaches only 400°C.

Space probes have shown some structure in the clouds, but the surface has only been glimpsed on a few occasions by the Soviet Venera space probes. They have managed to remain working for about an hour in the huge pressures, acid rain and extreme temperatures that exist on the surface. They show us a rocky surface of highly weathered rocks. From Earth we can only imagine such things. In the telescope Venus is almost featureless; it shows phases similar to the Moon, but little else. It is interesting to follow Venus through its phases and because of its plainness, it gives a good opportunity to start sketching.

Before photography was applied to astronomy, all observations made at the telescope had to be recorded by drawing. Today professional astronomers use highly sophisticated optical devices to record images, but simple sketches are often the best way for an amateur to make records of observations. To begin drawing you will need only a few pieces of equipment: a clipboard with a light, some paper or observing forms and a soft pencil and eraser.

You do not need to be an artist to make good observing records: with a little preparation there are a lot of short cuts which can make the job

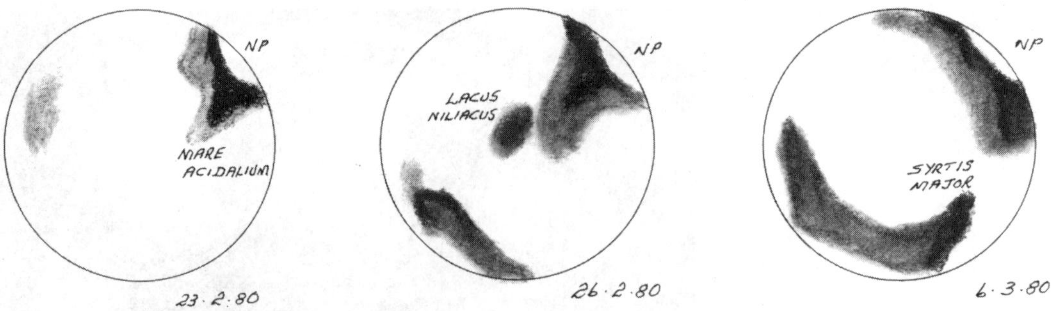

43 These observations of Mars were made by Ken Wallace during the opposition of 1980 using a 300mm telescope. Features visible in the drawings include Syrtis Major and the polar cap.

much simpler. A prepared observing form is a good idea. For planetary drawings, most of your work is already done for you, after all you know what shape the planet is going to be. For Venus, Mars, Uranus, Neptune and if necessary, Mercury, a simple circle drawn with a compass will suffice. Draw a square around the circle then use black ink to colour between the circle and the square: this will make the circle easier to use. Drawing outlines for Jupiter and Saturn and its rings are a little harder and will be described when we get to those planets.

The only thing you have to worry about is the size of the circles. If you make them too small then you won't be able to fit in all the detail, if you make them too large you will have trouble filling them. The best size is about 30mm to 60mm in diameter. You will need small blanks for objects on which you can't see much detail and larger blanks for those where you can see detail. It is a good idea to prepare a few master sheets and then photocopy duplicates for use.

You will need a clipboard or some other device for securing your paper. It will need to hold the paper at both the top and the bottom as you won't have any free hands. Make sure that you can illuminate the folder properly. You can either use your red torch and attach it to the board, or arrange for the light attached to the telescope or mount to be easily accessible for use. Another alternative is one of those small clip-on reading lights. If these are dimmed sufficiently with red cellophane, they are excellent for all sorts of jobs, sketching, reading maps and they can be made to hold the page open at the correct place.

When you are ready to sketch use a soft pencil, at least a 2B type, but I prefer a 4B; also make sure that you have a good quality soft eraser. If you can arrange to tie the pencil and rubber on to the board, so much the better: it saves crawling around in the dark when you drop one of them. When you begin sketching, fill in the large details first, the general positions of the features, then go back and fill in the details. Only fill in those features you are sure you have seen. If necessary change magnifications, increasing it if the seeing will allow, to get all the fine structure. Whenever you make a drawing always include a written description to clarify any obscure points. Make a note of any features that you glimpse but don't draw. Also record date, UT, seeing conditions and any other relevant information with your observations.

At first you will not be able to see any detail at all on the planets, but as you persevere with your observations you will begin to see more and more. Like many other fields of endeavour, the more you work at it, the more you get out of it. As you continue your observations you will begin to see detail quite clearly all the time and you'll wonder how you could possibly have missed it when you began. This is one advantage of keeping observing records: at a glance they will give you an account of the progress you have made.

With Venus, the most you will be able to see is the shape of the planet and the line separating the light hemisphere from the dark. Concentrate on getting these right, as it will help you in your future observations. Pay particular attention to the ends of the shadow line as sometimes bumps and hollows appear there; often these are optical illusions created by the contrast, but sometimes they are real features. Sometimes dark patches are seen on the illuminated part of the planet.

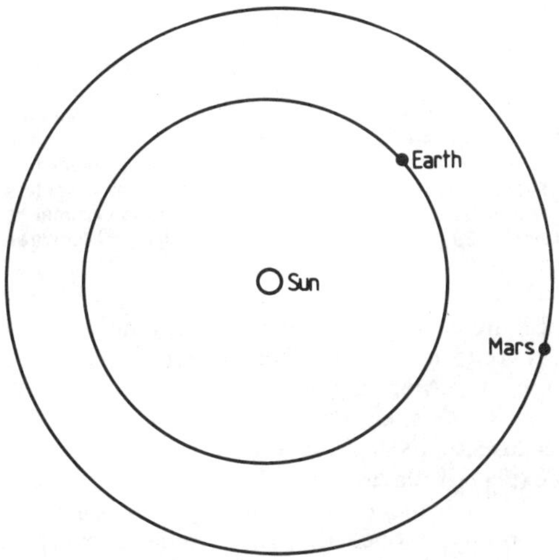

44 This diagram shows the paths that the Earth and Mars take in their journies around the Sun. As can be clearly seen, the distance between the two planets when they are at opposition can vary quite a bit.

Record any features you notice; they are interesting and if your records are ever compared with those of others they may be important.

Mars is the next closest planet to the Earth and the one on which you can start to see some detail. When scientists first started to study Mars with telescopes, they were able to see markings which appeared permanent. These markings appeared green against the predominantly orange planet. The markings were seen to fade at some times and to become much darker at others, so it was assumed that these patches were caused by vegetation. Large forests which varied in thickness due to the abundance or lack of water was one explanation. This idea quickly captured the imagination of the public, and soon the possibility of intelligent life on Mars was widely accepted.

In 1877 two important discoveries were made about Mars. The American astronomer Asaph Hall discovered the two small moons of Mars, Deimos and Phobos, and the Italian G.V. Schiaparelli discovered what he thought were linear features on the planet's surface: he called them 'canalli' meaning channels. Percival Lowell working in the United States called the features canals and said they must have been made by intelligent Martians trying to save water on their planet. Lowell was able to draw incredibly detailed maps of the canals, and to explain how they worked and how they affected the large regions of vegetation. The idea was disregarded by many astronomers who said that Lowell was deceiving himself, that the canals were an optical illusion. Recent visits by spacecraft have shown no sign that canals were ever present on Mars, but there are some features that do line up along some of Lowell's bigger canals, perhaps causing what looked to be a straight line.

The idea that life existed on Mars did not die with Lowell's canals. Scientists knew that the environment on Mars would not support animal life as we know it on Earth, the temperature is below freezing for most of the time, there is very little water in the air, most of it is trapped in the polar ice caps, and the atmosphere is very thin and almost all carbon dioxide. Despite all these factors against them, some very lowly plants, or bacteria, may have been able to survive in the warm—20°C at noon—temperatures of the equator. To test this theory, the United States sent two probes, the Vikings, to land on Mars. The photographs they sent back were spectacular, but the mini-laboratories that they carried were able to find no conclusive sign of any form of life. As far as we know, Mars is a dead planet.

Recently stories have appeared saying that there was once an ancient civilization on Mars. There is absolutely no scientific evidence that this is true. Photographs have been circulated that show what appears to be a large human face carved out of stone. This feature only appears as a face under special lighting conditions, under other illumination it looks just like what it is, a rocky outcrop. Romantic as the idea is, there are not and never were any Martians.

Observing Mars from Earth is a rewarding pastime; when the planet is at its best one of its features, a triangular region known as Syrtis Major, can be seen in even small telescopes. Unfortunately Mars is rarely at its best. To see a planet well you need to be as close as possible; this happens when the planet is at opposition. Opposition is when both the Earth and the planet are on the same side of the Sun and both are in a line with the Sun. When you are observing, this means that the planet and the Sun have to be directly opposite each other in the sky. The Earth travels around the Sun once a year, Mars travels around the Sun once every 687 days so the two are at opposition about every 780 days, a little over two years.

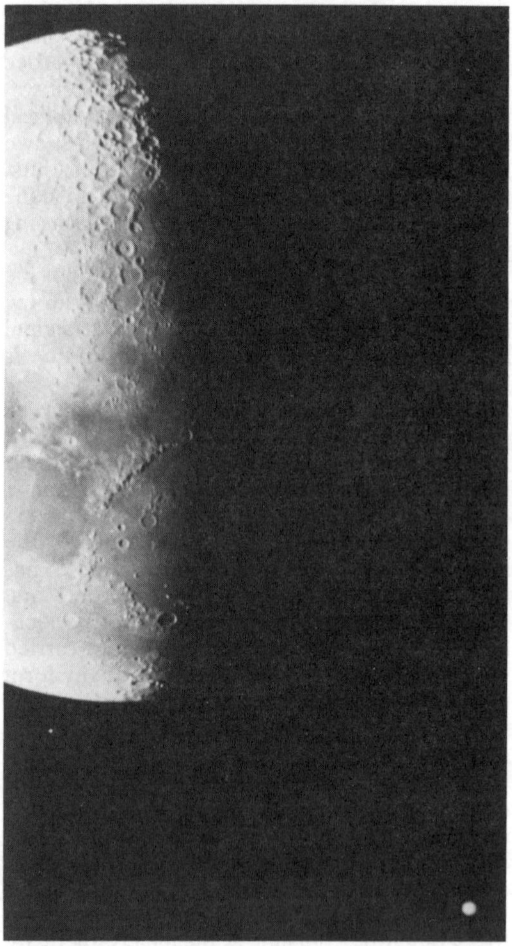

45 The Moon occasionally comes quite close to the planets. Here we see Jupiter just half a degree from the Moon's northern edge. PHOTO: KEN WALLACE

Another factor which affects how well you see the planet is how close we are to it at opposition. Because neither the Earth nor Mars have circular orbits, the distance between the planets at opposition can range from 100 million kilometres down to only 56 million kilometres at best. This will mean that Mars varies in size from 14″ to 25″ with an average of about 18″. Because of its small size, observations can only be made for a month or two either side of opposition. The list of coming oppositions is given in table 7.1.

Finding Mars in the sky when it is near opposition is easy: it will be at about magnitude −2.8, much brighter than any star; it also appears very red in colour. Even when it is not at opposition Mars is always above second magnitude, so it is easy to follow through the sky. Because of its high speed, it often pops up in unexpected places if you have not been observing for a while. When I was getting started in astronomy, I once thought I had discovered a new star in Scorpius. I checked all my maps and couldn't find it, I got the telescope out to observe it and found that it showed a disk; it was Mars. In the few months I hadn't been observing, it had moved a long way from where I remembered it.

Table 7.1 Coming oppositions of Mars

date	distance from Earth
28 September 1988	59 million kilometres
27 November 1990	77 million kilometres
7 January 1993	94 million kilometres
12 February 1995	101 million kilometres
17 March 1997	99 million kilometres
24 April 1999	87 million kilometres

Because there is detail to be seen on the surface, the selection of magnification is important. Too low a magnification will cause detail to be lost, too high a magnification will result in a poor image. The best way to get the correct magnification is to start with about medium power. If the seeing is good enough, you can increase the magnification, to try to find more detail, but remember the maximum limit. If the seeing is too poor to use medium power, then it is probably not worth doing serious observations of the planet for that observing session.

With a small telescope, there is little detail that is visible. Syrtis Major will show up as a dark patch on the planet and one of the polar caps may be visible as a white patch near the pole, but little else will be seen. To see any of the other features a telescope of about 200mm aperture is needed. Drawing Mars will require a prepared blank of about 30mm in size. Record any dark markings you see, and the position and size of the polar cap. Throughout an opposition, the size and shape of the polar cap may change, so it is a useful feature to record.

Mars has a rotation period very similar to that of the Earth, 24 hours 36 minutes. This rotation means that over an observing session you may well notice some movement of the features. If you observe at around the same time each evening you will see slightly different views of

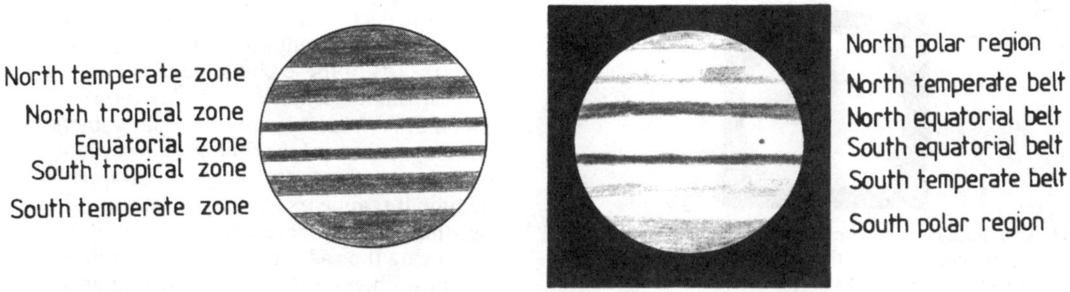

46 This observation of Jupiter shows the features visible, under excellent conditions, using a small telescope. The equatorial belts are clearly seen, the polar belts less so. The small dot on the equatorial zone is the shadow of one of the Galilean satellites as it passes in front of the planet. The bands of clouds around Jupiter are so constant that astronomers give them names to help identification of planetary features.

the planet. Over a period of a few weeks, all of the planet's surface will be presented for your viewing.

Jupiter is the next planet out from the Sun, and the largest in the solar system. It is so large that there is room inside it for more than 1300 planets the size of Earth; it is bigger than all the other planets combined. Although it never gets closer to us than 590 million kilometres, Jupiter appears larger than any planet except Venus, with an angular size of 47″. In the sky it reaches a brightness of magnitude −2.5.

Like Venus, the only thing we see on Jupiter is the tops of its clouds, the surface of the planet is hidden under tens of thousands of kilometres of atmosphere. The difference is in the structure of the clouds. On Venus we see an almost featureless surface, on Jupiter the clouds form coloured bands that circle the planet. These bands are so permanent that they have been given names to identify them. Jupiter's clouds contain features like the Red Spot: this is an anti-cyclone, larger than the Earth, which has been raging non-stop since it was discovered in 1631.

Observing Jupiter with even a small telescope is a fascinating experience. Each of the belts of cloud is coloured differently so it is simple to see at least some banding across the planet. Under very good conditions the red spot can be seen with a telescope as small as 50mm. Observing Jupiter is something you will need to practise. The more often you observe it, the more detail you will be able to see. Don't go out expecting to see features though: if you do you will probably see them whether they are there or not. Don't find out if the red spot is supposed to be visible, go and observe and if you think you see it, check in an ephemeris if it was visible after you have finished observing.

To identify features on the surface of Jupiter astronomers have a system of longitudes. Because Jupiter is mostly gas it rotates at different rates, the equatorial region turning faster than the polar regions. To make the system work astronomers have to use two longitude systems, System I for the equator and System II for the poles. Publications such as the Astronomical Society of New South Wales' Ephemeris, give you the system I and II longitudes for any time during the year. The red spot is in system II but it drifts around in longitude, so it is difficult to exactly predict where it will be.

Drawing Jupiter is worthwhile as there is so much detail you can record. The problem is that a simple circular blank is not much use for the planet. Because it is gaseous, Jupiter bulges quite a bit at the equator; this bulge is very noticeable. Blanks for Jupiter must be elliptical in shape. Drawing these takes a little patience; you will need a drawing board or some corrugated cardboard to hold your paper and into which you can stick pins. Take a piece of cotton and tie it so you have a loop 25mm long. Now take two pins and stick them into the board 5mm apart. Loop the cotton around the two pins and place your pencil inside the loop. Using the pencil to keep the loop taut, draw around. You will end up with an ellipse of about 23mm × 25mm. Use this to prepare an observing form which you can photocopy.

It is not only the surface of Jupiter that is interesting to watch. Jupiter has a collection of moons which are also clearly visible from the Earth. Four of these moons were one of the

47 An example of an observing blank for Jupiter. Using pins and a piece of thread it is a simple matter to construct your own blanks for use when observing.

earliest discoveries Galileo made with his first telescope; they have been called the Galilean moons in his honour. They are often numbered I, II, III and IV in order of the distance from Jupiter, but their names are Io, Europa, Ganymede and Callisto. All of these moons are bright enough to be seen with the naked eye from the Earth, except that the light of Jupiter itself swamps them. Through even the smallest telescope, the constant shuttling back and forth of the moons is an intriguing spectacle. Many publications include the positions of Jupiter's moons so that you can identify which moon it is that you are seeing. Jupiter has more satellites than just the Galilean four: there are another twelve in orbit around the planet, but none of these is brighter than thirteenth magnitude, so they can only be seen in fairly large telescopes.

In recent years unmanned spacecraft from the United States have visited Jupiter and sent back pictures of the Galilean satellites. Voyagers I and II have shown us that the worlds are very different from each other. Io is a world torn by tidal forces to such an extent that most of its surface is made of volcanoes. Europa on the other hand is the smoothest object in the solar system—its surface is covered by a crust of water ice with liquid water apparently underneath it. Ganymede and Callisto are cratered, but also covered by ice. In the next few years the United States plans to send more spacecraft to explore these fascinating worlds.

Through the telescope, the moons look like four bright stars strung about the planet. Sometimes it is possible to see all four moons on one side of the planet. At other times Jupiter hides one or more moons as they pass in front of, or behind it. Very rarely you can see Jupiter with all its moons obscured by the planet. As all the moons are seen in reflected light, sometimes they pass into each other's shadows; at these times one of the moons will wink out of view for a short period. Lists of these phenomena are provided in the Astronomical Society of New South Wales' Ephemeris. When viewing such phenomena, it is necessary to record the time very accurately. You must record the time the moon disappears and the time it reappears to the nearest second. Make sure that the timepiece you are using is accurate: set it using the radio time signals just before you start observing, and check the setting again after you have finished observing. Record the method you used to set the time along with the other records of your observations.

Another feature which was discovered when spacecraft reached Jupiter was that it had a faint ring of material around it. But it is not Jupiter that is famous for its rings, it is the next planet from the Sun, Saturn. Saturn is a smaller planet than Jupiter, but it is encircled by a spectacular system of rings nearly 300 000 kilometres in diameter. The rings of the planet can be seen in even the smallest telescope. Galileo saw them in 1610, but was unable to identify what they were because of the limitations of his lenses. For many years there was debate about what the rings were made from. Theories ranged from solid rings of rock to bands of small dust particles. We now know that the rings are made mainly from chunks of ice about a metre in diameter.

From Earth, the rings can be clearly seen as separate from the planet. Under good conditions and with a reasonably sized telescope, some structure can even be seen in them. There is a gap between two of the main ring systems, called the Casini division after the astronomer who discovered it. Sometimes this gap is visible as a dark band running through the rings. The rings themselves each have a different brightness, and this is often visible in a small telescope also.

As Saturn revolves about the Sun we see the rings from different angles, sometimes fairly widely spread, other times edge-on. When seen edge-on the rings appear to vanish for a few months. This last happened in 1980 and will happen again in 1995. During the time between these edge-on presentations we see the rings at different tilts. You will find that the rings are amongst the most popular items you can show people with your telescope. If you really want to impress people with astronomy, Saturn is a good place to start.

Saturn is also easy to find in the sky, it has a magnitude of −0.2 when it is near opposition, and appears dusky in colour. Like Jupiter, Saturn is a gas planet and shows cloud bands also. Because it is colder than Jupiter—noon temperatures are around −180°C—the colours of the bands and other features are not as pronounced as those of its bigger neighbour. The banding on Saturn is difficult to see with a small telescope under all but the best of conditions, so drawing the planet is of little value. If you do want to make some drawings, you will need to use an elliptical blank like that drawn for Jupiter, as Saturn is also noticeably flattened at the poles. Instead of colouring around the outside of the disk black, leave it white so that you can draw the rings in.

Saturn has a larger family of moons than Jupiter, but only one of them, Titan, is easily visible in a small telescope. Initial observations made Titan the largest moon in the solar system, but when spacecraft arrived in 1980 it was found that Titan has an atmosphere almost as thick as that of the Earth. This extra layer of gas makes Titan just a little smaller than Jupiter's largest moon, Ganymede. Watching Titan as it circles Saturn every sixteen days can be interesting, but the only time it passes in front of and behind the planet is when the rings are nearly edge on, at other times you can follow it through its entire orbit. Some of the other moons are visible in larger telescopes, but to find them you will need to follow the planet for a few nights to distinguish them from the background stars.

Out past Saturn are the last three planets. They are distant from the Sun and Earth, and none of them is a spectacular sight in even large telescopes. The first of these planets is Uranus. Although it wasn't discovered until the sixteenth century, Uranus is occasionally visible to the naked eye under excellent conditions, as it can reach magnitude 5.7 at a favourable opposition.

It was only in 1986 that we had our first close-up view of Uranus. Voyager II which had already seen Jupiter and Saturn arrived on 21 January to show us a new world. Like Jupiter and Saturn, Uranus has the banding of a large gas planet, although it is even fainter than Saturn's for the temperature at Uranus is about −190°C. Uranus' five moons, Titania, Oberon, Umbriel, Ariel and Miranda, were all photographed, giving us a look at some interesting places; but all are too faint for small telescopes, the brightest being Titania at magnitude 14.

To find Uranus, you will need to plot its position on a star map, and then use your telescope to look in the general direction. Put a low powered eyepiece in the telescope, because even at 40× the planet will show a noticeable disk. Search the area and when you find a disk with a greenish colour you will know you are looking at Uranus. You will not be able to see any detail on the disk, nor the very thin rings discovered about the planet, but you will have the satisfaction of seeing another of the Sun's family.

Past Uranus is Neptune, the last of the gas giant planets. Neptune has never been seen close up. Voyager II which visited Uranus is on its way and will arrive in August 1989. To find Neptune, get its position from an ephemeris and plot it on a star map. When it is at opposition Neptune has an angular diameter of 2.5″, so you will need at least 50x magnification to see it as a disk. At its best, Neptune has a magnitude of 7.6 so you should have little trouble finding it. Once you do see the planet you will notice it has a bluish colour. Neither of Neptune's two moons, Triton and Nereid, are visible in a small telescope as they have magnitudes 14 and 17 respectively.

Although it wasn't officially discovered until 1846, Neptune was first seen by Galileo, who didn't realise what it was. While he was observing the satellites of Jupiter, Galileo drew a star in the field of view; a few nights later when doing more observations he again drew the star in. Of course, Neptune had moved, and Galileo made a comment to that effect, but never took the observations any further. This shows the value of recording your observations, and occasionally checking them. Another interesting fact about Netpune is that it takes 165 years to travel once around the Sun; this means that we have not yet observed it throughout one full revolution. The first Neptunian anniversary of its discovery won't take place until 2011.

Well away from the rest of the solar system orbits the last of the planets, Pluto. Because of its immense distance and small brightness, we know very little about Pluto. At maximum brightness it reaches magnitude 15 so it is out of reach of any telescope smaller than 250mm aperture. It was only in 1978 that Pluto was discovered to have a moon, Charon. It appears that Charon is about one sixth the size of Pluto, making it huge

48 This one photograph shows three of the phenomena visible to the naked eye. The most obvious is Halley's comet during its 1986 apparition. To the right of the comet the long trail is due to a satellite passing across the sky. Just above the satellite trail is a small dash caused by a meteor as it entered the Earth's atmosphere.
PHOTO: STEPHEN BURNS

compared to the planet. Our own Moon is relatively the next largest. It is one eighty-first the size of the Earth.

Another odd thing about Pluto is its orbit. It is by far the most elliptical in the solar system. At its furthest Pluto is 7300 million kilometres from the Sun, at its closest, only 4400 million kilometres. This means that for about twenty years of its 248-year orbit Pluto is closer to the Sun than Neptune is. At the moment this is the case and Pluto will remain the eighth planet from the Sun until 1999. Pluto is certainly a strange world—it seems that we will know very little until we send a spacecraft there. Unfortunately there are no plans to do this in the near future.

Although there are only nine planets in our solar system there are many other objects that can be viewed. In chapter 2 we have seen that comets are also members of the Sun's family. A good comet is a large object in the sky and must be viewed with a low powered eyepiece for best effect. In the coma region near the nucleus of the comet you can sometimes make out detail, blobs and streamers of gas given off by the comet. This fine detail is perfect for drawings as it changes from hour to hour. When you are drawing the comet, your first sketch should be of the general area, mark the locations of the main features using a low powered eyepiece. Then, if seeing will allow, change up to a higher power to examine the finer detail and add it to your drawing.

A collection of drawings made over the apparition of a comet is quite a valuable scientific work. They will show the development of the comet while it is near the Sun. To make the drawings useful, make sure that you record all the details of the observations, including all the magnifications you used to record the detail. Once you have a record of an apparition, contact your local astronomical society as they can put you in touch with other amateurs and professionals who are also working on the comet.

The last of the Sun's family of objects are the asteroids. These are small rocky bodies which orbit the Sun, mostly in the region between Mars and Jupiter. There is a pattern in the distances of the planets from the Sun known as Bode's Law; this law predicts the presence of a planet in the gap between Jupiter and Mars. At the end of the eighteenth century a group of astronomers got together to find this missing planet. On 1 January 1801, the first day of the new century, G. Piazzi made an observation of an object that fitted the gap perfectly; he named it Ceres. Measurements showed Ceres to have a diameter of less than seven hundred kilometres, much too small for a proper planet, so the search continued. Between 1802 and 1808, three more objects, Pallas, Juno and Vesta were found, but no more were forthcoming. The 'planet hunters' stopped looking. It wasn't until 1845 that a German amateur discovered another; he called it Astraea. Everyone then started looking and since 1848 not a year has gone by without one or more asteroids, as they had become known, being discovered.

Although the four largest asteroids (the ones that were discovered first) are bright objects, none of them is visible with the naked eye—all of them are just below sixth magnitude; they are simple objects for amateur telescopes. You can find the positions of the asteroids from most ephemerides, and by plotting the position on a map, you should be able to find them. Because they are all small objects you will not be able to see any details, or even a disk. The asteroids appear as stars. To make sure that you have found an asteroid, make a quick sketch of all the stars in the field of view. The next night look again at the same position—one of the stars should have moved: that is the asteroid.

Although most of the asteroids are well behaved and stay between Mars and Jupiter, some of them have exceptional orbits. One asteroid, Icarus, has an orbit that can carry it closer to the Sun than the planet Mercury. Some get very close to the Earth—in 1937 a small asteroid just over a kilometre in size, Hermes, passed within 800 000 kilometres of us. With this in mind it is a valid question to wonder whether an asteroid could hit the Earth. The answer is yes, but the chances of it happening are incredibly small. Even if we were to find that an asteroid were heading for the Earth, there is nothing that could be done about it. The meteorites which impact upon the Earth every so often are probably small chunks of material from the asteroid belt.

Watching the smallest members of the solar system can be fun; it can also be profitable. If you discover a minor planet and it is recognised that you found it first, you have the honour of naming it. It was traditional to name asteroids after figures from mythology, but with many thousands now known, mythological names ran out. In recent years it has become commonplace to sell the naming rights. There are now some asteroids with weird and wonderful names though most are much too faint to be seen with all but the largest of telescopes.

8

Our Sun and Moon

The two brightest objects in the sky are the Sun and the Moon—because they are so bright they are the easiest of all objects to view. We will discuss the Moon first. It is certainly the best object to start observing with and even with only a small aperture, there is an incredible amount of detail to be seen.

The Moon is Earth's nearest neighbour, being just 400 000 kilometres from us. It was the first object reached by a space craft: the Soviet-made Luna 1 passed within 7500 kilometres in January 1959. On 14 September 1959 we made our first impression on the Moon when Luna 2 crash landed on it. The Moon was also the first body other than the Earth that man walked on, with the United States' Apollo XI landing on 21 July 1969. The last people to walk on the Moon did so in December 1972.

Apart from being a great achievement, man's trips to the Moon brought back many rock samples to tell us exactly what the Moon is made of. Before the trips the theories about the Moon varied from the scientific to the ridiculous. In the 1800s there were many reports of life on the Moon, stories of advanced civilizations of 'Moon maidens'. Today we know the facts; the Moon is a completely dead planet with no atmosphere and no water except for a little trapped in rock crystals. With daytime temperatures of nearly 100°C and night temperatures of −150°C it is difficult to think of a more inhospitable place. The rocks of the Moon have not revealed any sign of life either now or at any time in the past. The Moon maidens and advanced civilizations have gone the way of the canals of Mars.

The rocks that were brought back to Earth by the Apollo astronauts and unmanned Soviet spacecraft have shown us that the Moon is made of roughly the same materials as the Earth. The rocks are similar to basalts and breccias on the Earth, the main difference is the amount of water locked in the crystals of the rocks. There is a fair bit of moisture in terrestrial rocks, but very little in the lunar ones. the Moon has always been a very dry place.

The Moon was formed at the same time as the Earth, and by practically the same processes. The rocky material that was left over from the formation of the Sun gradually collected together to form the rocky planets, Mercury, Venus, Earth, the Moon and Mars. The difference between the Earth and the Moon is simply that the Earth collected more matter. The Moon has a mass about one eighty-first that of the Earth. This mass difference means that the Earth was able to retain its atmosphere, to keep water as a liquid and to support life. The Moon, because of its lower gravity, lost its atmosphere and water.

Throughout their histories both the Earth and the Moon have been bombarded by meteorites and astronomical debris. Because of the weather on the Earth and the erosion it causes, most of the traces of these collisions have been erased. We are left with just a few marks such as the Barringer crater in Arizona. The Moon retains its scars, so when we look at it we can see the results of this continual bombardment. This long history of impact cratering has left a number of features that can be seen from the Earth even with only a pair of binoculars.

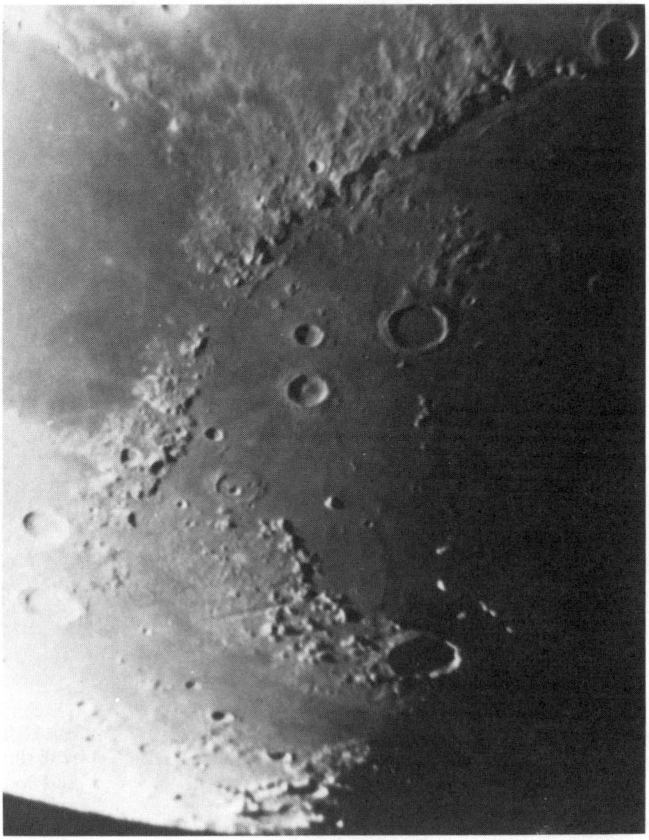

49 The Mare Imbrium is a good example of a mare region of the Moon. With the lunar alps at the bottom of the photo and the alpine valley running through them, this area is one of the most interesting on the Moon. PHOTO: KEN WALLACE

The first of the features that can be seen are the large dark patches over the northern hemisphere. Before the advent of the telescope astronomers thought that these dark patches were oceans and seas and called them romantically, names such as Sinus Iridium (Bay of Rainbows) or Mare Tranquillitatis (Sea of Tranquillity). Of course the ancient astronomers named them all using Latin, and this is how they are still officially known today. Who said it was a dead language? Table 8.1 gives the translations. Although they are now known to be old plains of lava, the romantic names have remained and the features are still called seas.

The seas are relatively new features on the planet compared with the mountains of the highlands. The seas formed about 4000 million years ago when huge meteoritic impacts caused the surface to open and great flows of lava to spew forth. These lakes of molten rock solidified, giving us the relatively smooth surface seen there today. In many places these lava plains have joined together, but in others the remains of the crater walls can still be seen. The south-eastern portion of Mare Imbrium is a good example, as are the northern shores of Sinus Iridium.

After these plains had solidified they were still bombarded by meteorites and some of the best craters are visible in the middle of these seas. The crater Copernicus is in the Oceanus Procellarum and is one of the best craters on the Moon. It is over eighty kilometres across with walls around it rising over five kilometres from the lunar surface. If you look carefully at the centre of the crater you will notice a small mountain. These central peaks are created when the crater forms and the Moon's surface starts to return to shape. Other craters such as Archimedes in the Mare Imbrium have no such central rise.

Craters are scattered all over the Moon, but nowhere as heavily as the highland regions of the southern hemisphere. This area of the Moon is the oldest, so it has had longer to be hit, hence the heavy cratering. In this area you will notice a funny thing: many of the craters form chains, three, four or more large craters that seem to line up. In the early days this was thought to be evidence that the Moon's craters were predominantly caused by volcanic action within the Moon rather than meteoritic impact. Today it is thought that these chains are simply a statistical feature; there are so many craters that some line up just by chance.

One of the features of the highlands is the great walled plain of Clavius. Clavius has a diameter of 233km and an area greater than that of Switzerland. Within Clavius there are a number of smaller craters. Some of these are in the flat central area but there are two which break through the plain's wall. This is common on the Moon, quite often the floors and walls of large craters are marked by smaller craters. Selenologists, as those that study the Moon are called, have taken this to indicate that the size of the meteors hitting the Moon got smaller as time went on. Nowadays there is no more cratering going on, because objects large enough to cause a visible crater are extremely rare in our region of the solar system. The last report of a crater forming for which there is any hard evidence, and even that is not conclusive, is the case of Messier A.

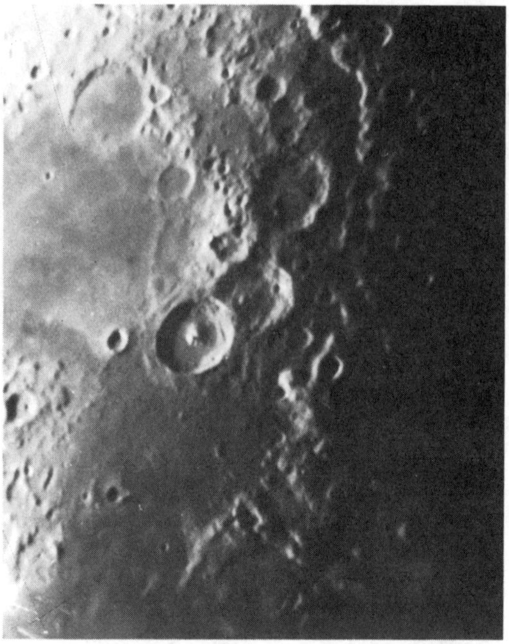

50 This area of the Moon's eastern side near the Mare Nectaris shows typical features of cratered regions. Three large craters, Theophilus, Cyrillus and Catharina can be seen near the middle of the photograph along with the crater Fracastorius which had its wall breached and floor filled with lava when the mare formed. PHOTO: KEN WALLACE

Table 8.1 The 'oceans' of the Moon

Latin Name	English Translation
Mare Australe	Southern Sea
Mare Crisium	Sea of Crises
Palus Epidemiarum	Marsh of Disease
Mare Foecunditatis	Sea of Fertility
Mare Frigoris	Sea of Cold
Mare Humboldtianum	Humboldt's Sea
Mare Humorum	Sea of Moistures
Mare Imbrium	Sea of Showers
Sinus Iridium	Bay of Rainbows
Mare Marginis	Marginal Sea
Sinus Medii	Central Bay
Lacus Mortis	Lake of Death
Palus Nebularum	Marsh of Mists
Mare Nectaris	Sea of Nectar
Mare Nubium	Sea of Clouds
Mare Orientale	Eastern Sea
Oceanus Procellarum	Ocean of Storms
Palus Putredinis	Marsh of Decay
Sinus Roris	Bay of Dews
Mare Serenitatis	Sea of Serenity
Sinus Aestuum	Seething Bay
Mare Smythii	Smyth's Sea
Palus Somnii	Marsh of Sleep
Lacus Somniorum	Lake of Sleepers
Mare Spumans	Foaming Sea
Mare Tranquillitatis	Sea of Tranquillity
Mare Undarum	Sea of Waves

Two of the early pioneers of selenography, Wilhelm Beer and Johann Madler drew and described two craters in Mare Foecunditatis, Messier and Messier A, as being identical. Today Messier and Messier A appear completely different; they have different sizes and shapes. If Beer and Madler and their three hundred drawings are correct, then something has happened to Messier A since 1837, probably the impact of a small meteorite.

Another feature of the highlands region is the most conspicuous object on the full Moon, the crater Tycho. This crater, named after the famed astronomer Tycho Brahe, is only 86 kilometres across but it stands at the centre of a system of rays that stretch across a quarter of the Moon. When the object which formed Tycho hit the

51 This photo of the full Moon shows the rays of light material associated with the crater Tycho quite clearly. Note that because of the angle of illumination, very few other craters are visible on the Moon's surface. PHOTO: KEN WALLACE

Moon it must have sprayed up a huge amount of light coloured rock from below the surface. This rock flew out creating an extensive system of rays all pointing away from the crater. Because they are so intriguing, the United States sent one of it Surveyor spacecraft, number 7, to investigate. It landed on the slopes of Tycho and found that the rays had an abundance of aluminium and relatively little iron compared with other Moon rocks. Under shallow illumination, such as that around last quarter, Tycho doesn't look any different to the thousands of other craters in the region, but at full Moon the rays are so prominent that there is no way that you can miss them.

Craters and lava seas are not the only features to see on the Moon, though most of the others require fairly large telescopes to be visible. One feature that can be seen in a 50mm telescope is the straight wall. This is a feature that lies in the south-eastern part of the Mare Nubium and although it is neither straight nor a wall, it is an intriguing feature. This line on the Moon's surface is in fact an ancient fault in the rocks. It runs for nearly 110 kilometres and is about 300 metres tall in places. Before full Moon, the wall appears as a dark line running across the plains; after full Moon it appears, not as distinctly, as a bright line.

Because most of the Moon's surface is the same dull grey colour, the only way we can observe most of its features is by looking for the contrast between a feature and its dark shadows. The only source of light the Moon has is the Sun, so it is the position of the Sun that determines what we will see when we observe. The most obvious effect the Sun has is to cause the phases of the Moon. Through the month the phases change: a thin sliver of light appears along the eastern limb of the Moon—this moves westward, so after a week the whole eastern half of the Moon is illuminated. During the next seven days, more and more of the Moon becomes illuminated until the Moon is full. Over the next fortnight the Sun moves westwards and less of the Moon is illuminated, until just a small amount of the western limb can be seen. The cycle then starts over again.

The best time to observe any feature on the Moon is when it is near the terminator, the line separating the light and dark parts. The terminator marks dawn or sunset on the Moon. Like the Earth, this is the time that the shadows are

52 The Moon a few days after New. Craters and other features can clearly be seen along the terminator in this 1/15 second shot. PHOTO: KEN WALLACE

longest and the view is best. By concentrating on the area near the terminator there will be new features to view each night. Over a month you will get a good idea of the interesting features that can be seen. One difficulty with following the Moon through a month is that it rises about fifty minutes later each day, so while first quarter Moon is convenient, rising at midday, last quarter doesn't make an appearance until midnight, and the later phases are even worse. To observe the last part of the Moon's cycle you will need to rise early in the morning.

At first glance there appears to be more detail on the Moon than you can possibly comprehend, but by taking it slowly you gradually learn your way around. When observing the Moon it is best to start with a fairly low powered eyepiece. The Moon is a very large object, being over half a degree in diameter, so it is often a problem to fit it all into view. Low power will give you a good, bright image, then you can scan over the surface looking for interesting features to observe more closely. Once you have had a look at the overall view it is time to change to a medium power.

With a medium powered eyepiece in position, you will be able to see selected features close up. Move the view to the area along the terminator, and then move your scope up and down to view each part. If you observe for more than an hour or so, you will easily notice the motion of the terminator across the area you are looking at. Because of the abundance of light, the Moon is one object on which you can use a fairly high power, but be careful that the seeing conditions will allow this.

If you have decided to try drawing your observations, there is hardly a better place to start than with the Moon. Not only is there a huge amount of material from which to choose, but because it is so bright, you can have a much higher amount of illumination than usual to help you see what you are doing. When you're getting started you can remove the red cellophane from your torch and not have this affect your observations. Instead of a circle it is better to use a 50 mm-sided square to draw in. You don't have to use all the square, but it's a little easier than trying to fit things into a circle.

To get started with drawing, choose a simple feature, a crater on one of the maria is a better place to start than a mountain in the highlands. The first thing to do is to get the shape of the crater right. Although most of the Moon's craters are in fact circular, very few of them appear that way from the Earth. We usually see them foreshortened by the Moon's curvature, so these features end up being roughly elliptical in shape. Once you've got the outline the correct shape, you can concentrate on getting the rest of the crater right. Look for smaller craters within the large one, perhaps breaking through the walls in some places, and any central peaks, or peaks in the crater's walls and ramparts. The floor of a crater often has a fair bit of detail in it; be sure to record this. By starting with simple tasks, you

53 The Moon at first quarter. Taken only a few days after figure 52 it is difficult to pick out the craters seen in that photograph. PHOTO: KEN WALLACE

will very quickly progress to more complex ones, the highland regions, or the very complex regions near the Moon's edges.

The materials to use when sketching are similar to those used for planetary work, a clipboard, soft pencil and eraser, but if you are artistic you may like to try some other media. When you have finished a drawing there is nothing wrong with touching it up, as long as you don't add or alter any detail. Remove stray pencil marks, perhaps make the shadows darker; it is pemissible to only sketch the outline of the shadows and to fill them later with black ink. There is no reason why your drawings, apart from being accurate scientific records, shouldn't also be works of art.

There is so much to see on the Moon, too much to be described fully here, but I will include some parts of the Moon that are my favourites. Just as a new lunar cycle starts the terminator moves across Mare Crisium in the northern hemisphere. From the Earth this large plain appears as an egg-shaped patch, but Luna I showed it to be beautifully circular and surrounded on all sides by mountains and craters. All around this area, and in the southern hemisphere, there are many large craters, all of them have been stretched into long narrow shapes by foreshortening. As dawn creeps across this area of the Moon it is not uncommon to see the floor of a crater still in darkness while its edges and central peak are picked out by the early morning sunlight.

As the terminator moves westward, we see Mare Tranquillitatis, the site where man first walked on the Moon. The exact place is near to the south-west edge of the sea, and although there is nothing visible from Earth, it is nice to be able to see this historic site. Other parts of the mare show small deep craters which are also interesting to look at. Again look for central peaks, there are some high ones near the northern edge of the sea.

As we reach first quarter, some of the very large craters become visible. A little south of the equator are the 'big three', Ptolemaeus, Alphonsus and Arzachel. These three are all nicely lined up and make a wonderful sight as the sunlight reaches them. Just below the big three are three more large craters lined up. Just to the west of the gap between these crater chains is the straight wall mentioned earlier. As the Sun moves further, the crater Tycho becomes visible. It is about two thirds of the way from the equator to the south pole, but don't be concerned if you can't find it now, you'll have no difficulty at full Moon. Even closer to the pole is the large walled plane of Clavius. This is a fascinating sight, with all the smaller craters littering its floor. Clavius is also famous for being the site of the Moon base in the film '2001'.

At the north of the Moon is the Mare Imbrium and the largest feature on the Moon, the Oceanus Procellarum. In Procellarum is the large crater of Copernicus with its large central peak. Just to the north of this crater and separating the ocean and the sea, are the Carpathian Mountains. This mountain chain contains peaks which reach over two kilometres above the plains. On the north-western edge of the Mare Imbrium is the Sinus Iridium. This feature is a large crater that was overrun by lava when the mare formed, leaving only its northern and western walls.

54 This observation of the area around the Sinus Iridium region of the Moon was made on 10 January 1987 using only a 50mm refracting telescope and the clipboard system shown in figure 40.

Where the two features meet you can just make out the remains of the old walls.

As full Moon approaches, more of the Oceanus Procellarum becomes visible. The last features to come into view are the craters Grimaldi and Riccioli. These are so foreshortened that they appear as little more than dark blotches on the surface. At full Moon the shadows have disappeared and there is little to see, except of course for Tycho. After full Moon, the terminator again moves across the surface; this time however the shadows are in the opposite direction and many of the features take on new appearances; it is just as interesting as the first half of the show, even if the hour is a little inconvenient.

A type of observation that can be made even by beginners is occultation timing. An occultation occurs when one astronomical body passes in front of another, blocking it from view. The Moon in its monthly passage around the sky often passes in front of stars, occulting them. By accurately timing such occultations, and by using observations from all over the world, astronomers can refine their calculations of the size and orbit of the Moon. Predictions for these events are made by the International Lunar Occultation Centre in Tokyo. Local ephemerides often reproduce listings of events for the area the publication covers. There are two types of events to watch, disappearances and reappearances. Disappearances are easier to time as you can watch the Moon nearing the star; reappearances involve watching the limb of the Moon waiting for the star to pop into view. Most occultations are viewed against the dark limb of the Moon, but occasionally for a very bright star or a planet, bright side predictions are given.

When timing these events you must be as accurate as possible; to do this to the required precision you will need a short-wave receiver and a stopwatch. There is a service of radio time signals known as VNG; these are a series of short beeps each second, with a longer beep to mark the beginning of a minute. Using these time signals, start the stopwatch at the beginning of a minute. Use the beeps to check your accuracy, with a little practice it is possible to start right on the dot. You must set the stopwatch as close to the time of the occultation as possible, as this will reduce any error from the stopwatch. As soon as the star disappears or reappears, stop the watch and record the time. It will take you a certain length of time to react and stop the watch, but you should not worry about this as it can be allowed for when the observation is used. Any observations that you make can be passed on to your local society if they have an occultation section, or you can send them directly to Tokyo.

A much more spectacular event is a grazing occultation; this occurs when the Moon passes very close to a star. When this happens the star is seen to wink in and out of view as it passes behind the mountains and valleys of the edge of the Moon. Grazing occultations are very useful to astronomers, as not only do they give information about the Moon's position, but they also

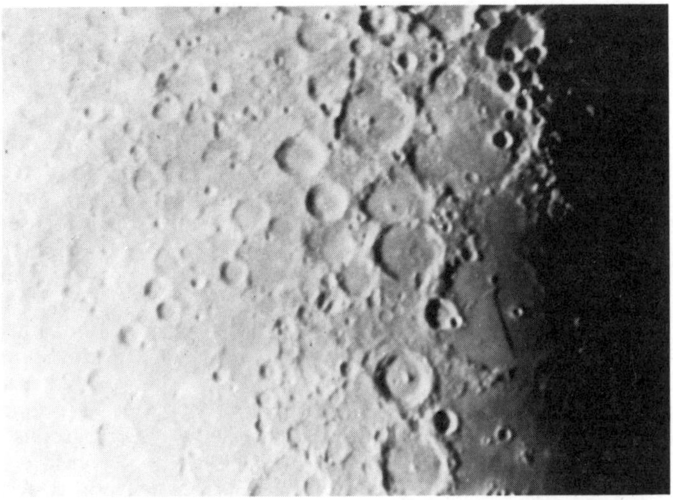

55 The famous straight wall at the edge of the Mare Nubium can be clearly seen in this half second exposure made with a 200mm telescope and eyepiece projection.
PHOTO: KEN WALLACE

allow accurate mapping of that region of the lunar surface. Besides, they are fun to watch.

The grazing occultation is only visible from a strip of the Earth perhaps a kilometre wide, so making sure you are at the right spot is very important. Many astronomical societies organise trips to watch grazing occultations, spreading their members out so that everyone gets slightly different results and a map of the area can be drawn. To observe correctly, you will need your short-wave radio again, but also a tape recorder. Because the star will come in and out of view many times during a good graze it is too difficult to time each event individually. A better way is to record the VNG time signal on the tape along with your voice. At the start of the tape you record the time to the nearest minute, then let VNG do the rest. As the star winks in and out, you speak into the tape recorder saying 'here', 'gone' or 'in', 'out' to identify what is happening to the star. When the event is over you can take your tape and make the timings at your leisure. Be sure that you use fresh batteries in your tape recorder, otherwise the tape may run slow making it difficult to make timings later.

The Sun is the closest star to the Earth, and the source of all life on the planet. All the energy we use, apart from nuclear energy, comes from the Sun. Plants catch sunlight and turn it into sugars, we eat the plants, or we eat animals that have eaten plants to get our energy. We burn wood, or coal and oil—the remains of long dead plants and animals—for fuel. Apart from being so important to life, the Sun is also important to astronomy—as the closest star to us, it is the only one on which we can see any detail. In chapter 9 we will look at stars in general, but many of our theories about them stem from observations made of the Sun.

The Sun supplies us with a great amount of light and energy, so much in fact that it is difficult to observe the Sun in safety. Looking at the Sun for more than a moment will cause damage to the sensitive retina in your eye. With optical aids gathering even more light and concentrating it, just a glimpse is sufficient to cause blindness. *Never*, under any circumstances, look at the Sun's surface with any optical instrument. Also never use the solar filters that come with some telescopes; they are simply unsafe. There is a simple, safe method to view the Sun and all its features, the solar projection screen.

To make a solar projection screen you will need a few things in addition to your telescope. You will need a non-cemented eyepiece such as a Huyghenian. Because of the amount of light and heat passing through them from the Sun the glue used in many eyepieces will melt or discolour. You will also need some cardboard or a cardboard box, paper and a star diagonal if you have one. The magnification you use is also important, the Sun has the same angular size as the Moon, so you will need a low power to fit it all in. It is best to stay below 50×, the projection device itself will enlarge the image sufficiently for you to get a good view without high magnification.

56 The crater Copernicus in the Oceanus Procellarum is shown in this photograph. Notice the fine central peak in this formation and the other less distinct features in the surrounding area. PHOTO: KEN WALLACE

To construct a projection device, fit the eyepiece to the diagonal or straight into the drawtube. The problem is to arrange a screen about 50cm from the eyepiece. With a Newtonian you can deflect the sunlight down the outside of the tube and onto a screen attached to the side of the telescope. Another screen in front of the eyepiece holder will be needed to block the direct sunlight falling on the screen. With a refractor, arrangements can be made for the image to be projected towards the ground. A cardboard box can then be used to catch the view. You will need to be able to tilt the box and adjust its height so that the image always falls straight into it. Figures 58 and 59 show you some arrangements you can use for solar observing. None of them takes more than a few minutes to construct, but they are quite suitable for the work. It is possible to buy solar projection screens for refractors, but these are expensive and really no better than the home made version, they just look nicer.

When you have your projection device assembled, take the telescope out and point it at the Sun. When you are doing this, don't use the finder scope or any other method which involves looking at the Sun. Look at the shadow cast by your telescope. When this shadow shows only the circular shape of the telescope tube, the scope will be pointing straight at the Sun. Hold a piece of paper near the eyepiece, move the telescope around slightly until you see a very bright dot appear on the paper; this is the Sun's image. Now set up the projection screen if you haven't already done so and focus the image.

When focusing the image it is important to make sure that what you are looking at are really features. Any dust in the telescope or its eyepiece will show up as dark spots on the image unless it is focused precisely. To check that you really are seeing the Sun's features, give the telescope a gentle nudge. If the spots move against the Sun's image, they are dust particles. Keep focusing until the only things visible as spots move with the Sun's image.

The surface of the Sun that you see is not solid, it is simply the top of a layer of gas. The Sun is gaseous all the way through, from the 6000°C surface to the 20 million degree core where the nuclear reactions which power the Sun take place. The layer that we see is called the photosphere, above it is the corona, a region of rarefied gases that have a temperature of millions of degrees. Under normal circumstances this layer is invisible from Earth, but during a total solar eclipse it can be seen and measured.

When you look at the image you will see that the Sun is not the featureless disk you probably expected; it has marks on it. The small black patches you can see are sunspots, places where the surface temperature of the Sun is lower, and so not as much light is given off. Seen by itself a sunspot would glow very brightly, but compared with the much greater brightness of the Sun's normal surface they appear dark. If you look carefully at one of these spots, you may notice that it is made up of a number of smaller sunspots. Figure 60 shows some normal sunspot groupings. Scientists have found that sunspots are associated with the Sun's magnetic field. Where the field breaks through the surface of the Sun it cools the gases, causing a sunspot.

If you were to count the number of sunspots over a reasonable period of time, you would notice a funny thing happening: the numbers visible on any day change. In some years there are sixty or so spots visible each day, in other years there are only one or two. This change in numbers follows a regular pattern of eleven years and is known as the sunspot cycle. Another change that follows the sunspot cycle is the position of the spots. Near the beginning of a cycle, spots appear at latitudes of about 60°. As the cycle continues, they move closer to the Sun's

57 Clavius is one of the most prominent of the features in the Moon's southern highlands. Here we see it foreshortened by its proximity to the Moon's southern limb. The crater Tycho, so noticeable at full Moon is seen close to the bottom of the photo. Because of the lighting however, it doesn't stand out. PHOTO: KEN WALLACE

equator; the cycle then repeats with more spots forming at high latitudes. When the positions of the spots are plotted on a graph (figure 61 shows this) it is hardly surprising that it is called a butterfly diagram.

A useful task you can set for yourself is to count and map the positions of sunspots each day. To do this prepare some observing blanks. On a piece of paper draw a circle that is the same size as the image of the Sun on your projection screen, then photocopy some duplicates for use. To make observations, you need only attach one of your prepared blanks to the observing screen and trace the positions of the spots. If there are any complex spots, you can also draw a quick freehand sketch of them to add detail to your observations. Make sure that you include all the usual items, time, date and so on, with your observations.

Before removing the sheet from the screen you will need to work out the orientation of the Sun. Without moving the telescope, let the image of the Sun drift across the paper: the direction it moves towards is west, the direction it moves from is east. To work out north and south, you will need to know how many times the light has been reflected, whether this is an even number or an odd one. A refractor with no star diagonal doesn't reflect the light at all, so count it as even; a Newtonian with a star diagonal reflects it three times, once at the main mirror, once at the secondary then once at the diagonal. Once you know the number of reflections, use figure 62 to work out north and south. A record of the

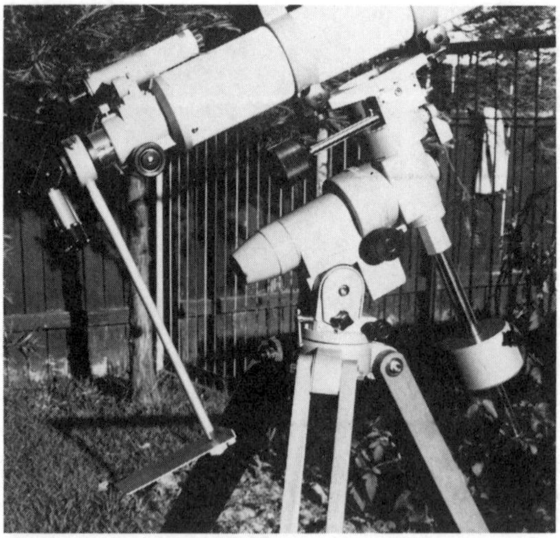

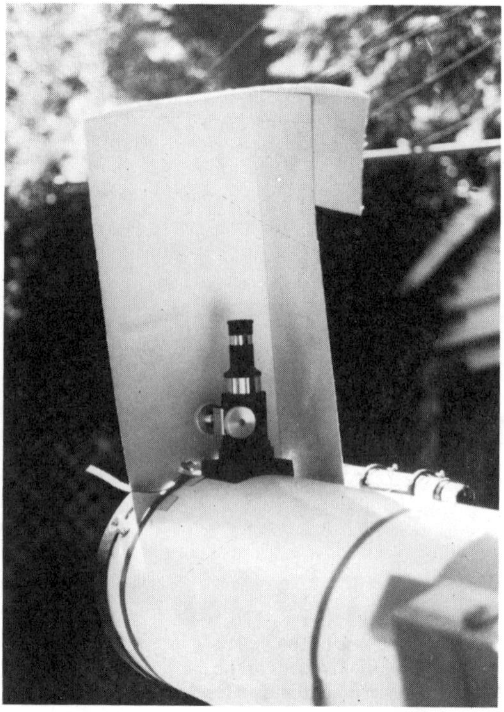

58, 59 These two photographs show two methods of solar projection devices. The one above is a commercial unit designed to fit onto the drawtube of a refracting telescope. The one opposite is made from cardboard and is designed to attach to the author's 200mm Newtonian with a plastic binding strap. PHOTO: DAVID REIDY

sunspots over a period of years can be a very useful project. Most astronomical societies have a solar section of some sort and many of these run observing programs.

There are many other features on the Sun, but they can't be seen without very expensive equipment. The hydrogen-α filter is one accessory that some well-heeled amateur astronomers use. These filters, which cost thousands of dollars, allow you to view the Sun in a very narrow band of light, around the colour given off by hydrogen. Using one of these filters, features such as solar flares and prominences become visible, huge columns and loops of gas, many times larger than the Earth, spewing out into space. Some astronomical clubs invest in one of these filters for use by members.

The Sun is a fascinating object, one that you can make a career of studying. It allows the amateur to make a real contribution to astronomy by observing, drawing and counting sunspots, as it presents its ever changing face to us. The Sun is also involved in some of the most spectacular events in astronomy as we shall see.

Like every other object, both the Earth and the Moon cast shadows in the sunlight. These shadows form cones that spread out behind the two objects into space. Because the Sun has a size, there are two parts to the shadows it casts. You can observe the two parts by looking at the shadow of a large building on a sunny day. If you were to walk into the shadow, you would enter a region in which the Sun was only partially obscured by the building. On the ground this would appear as a dimming of the sunlight. As you continued into the shadow, eventually the Sun would be completely hidden, and the shadow would be total. These two types of shadows have names, the umbra for the total shadow and penumbra for the partial shadow. Behind the Earth, the penumbra extends indefinitely, but the umbra comes to an end after about 1.3 million kilometres. Figure 64 shows this.

As our shadow stretches out into space it is possible for other astronomical objects to pass into it. The outer planets of the solar system can pass into our penumbra, but this has no observable effect on them. The Moon is close enough to pass into both the umbra and penumbra; when this happens, we call it a lunar eclipse. A little

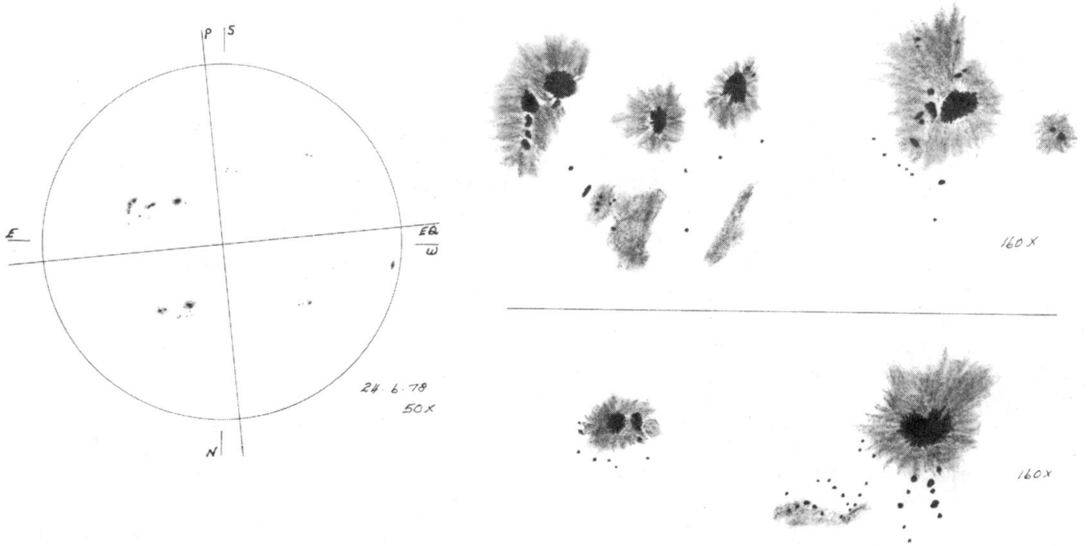

60 Observations of the Sun made on 24 June 1978 from Sydney. On the Sun's disk are two sunspot groups and a number of smaller spots. The close up drawings show details of the groups.

thought will show that this can happen only when the Moon is opposite the Sun in the sky; this happens every full Moon. The question is, why don't we have an eclipse at every full moon?

The reason is easy to see. Although the Moon and the Sun both seem to travel around the Earth, they do so on slightly different paths. The Sun, as we have seen, follows the ecliptic; the Moon however follows a path at a small tilt to the ecliptic. The Moon's path crosses the ecliptic at two places called nodes. When a full Moon occurs near one of these nodes, we have an eclipse. Sometimes full moon occurs when the Moon is close to a node, but not close enough to enter the umbra completely. It may enter just a part of the umbra or maybe only the penumbra. In either case we call it a partial eclipse. If the Moon does make it completely into the umbra, it's a total eclipse.

Observing an eclipse of the Moon is a very simple task, you can do it without a telescope, but with one you can do some useful astronomy. As the eclipse starts, the Moon begins to enter the Earth's penumbra. This is called the 'first contact'. At this time you will be able to see a dimming of the Moon's light, but with the telescope it is difficult to pinpoint the exact location of the Earth's shadow. A little while later the Moon will begin to enter the umbra: this is called the second contact.

The umbra's edge is easy to see both with a telescope and without. With a telescope you can follow the edge as it enters craters and crosses features. There are observing programmes that you can take part in which involve recording the times that the shadow's edge enters and leaves craters. Your local astronomical society can put you in contact with those running such programmes. The programme will have an observing sheet listing objects to be viewed during the eclipse. To make these observations of the eclipse you will have to make some preparations. You will need a large Moon map and some numbered stickers.

The first thing to do is to select those objects you will observe; you don't need to time all those objects on the list, just be content to watch those features that you can be sure of identifying. Once you have picked a few craters mark their positions on your Moon map with a numbered sticker. Keep the map well illuminated when you are observing. As the shadow moves across a feature, mark down its number and the time to the nearest second. Make sure that the clock you are using is accurate. I remember trying to time the lunar eclipse of 30 December 1982. I spent a

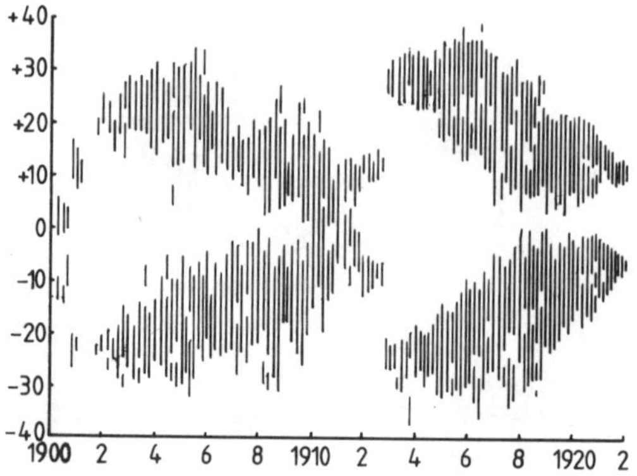

61 A graph of the latitudes of sunspots observed from 1900 to 1922. When plotted in this manner it is not hard to see where the name 'butterfly diagram' came from.

few nights making sure I could find all the craters I wanted to time, got a complete set of NASA maps ready, and was rained out. Another example of Murphy's Law in astronomy: the amount of cloud cover is proportional to the rarity of the event.

After a while the shadow will be completely across the Moon; this is called third contact; now is the time you can relax for a while and enjoy the view. The first thing you will notice is that although the Moon is completely within the Earth's shadow you can still see it. The layer of air around the Earth acts like a lens, bending some of the light from the Sun into the umbra—it is this that illuminates the Moon. Because of its passage through our atmosphere the light reaching the Moon is highly coloured, usually a dark red. Sometimes the colour will be a light orange, at others almost no light will get through. The darkest eclipses are always just after there have been large volcanic eruptions, such as those after Krakatoa's eruption in 1884.

As the Moon begins to leave the umbra, light begins to reach the surface again—this is fourth contact. As the edge of the shadow again crosses the marked features you will need to mark the times again. Eventually the umbra completely leaves the Moon's surface, the fifth contact. When the Moon completely leaves the penumbra we have last contact and the eclipse is over. The whole process will take up to five hours from first to last contact, with the time spent in the umbra anything from a few minutes to two hours. Even if you do not make crater timings, a lunar eclipse is a fascinating thing to watch. The movement of the shadow across the lunar surface will keep you entertained for hours.

Far more spectacular than a lunar eclipse is a solar eclipse. We have seen that the Sun, Moon and Earth line up best when the Moon is near one of the nodes of its orbit; if this happens at new Moon rather than full Moon, the Moon passes across the face of the Sun. Simply by coincidence the Moon appears to be about the same size as the Sun in our skies; because of this it covers the solar disk just nicely when seen from here. What is happening to observers on Earth is that they are moving into the shadow of the Moon, its penumbra and umbra.

Because of the Moon's small size, its umbra extends only about 400 000km into space, just reaching the Earth's surface, so only a small area of the Earth's surface will see the Sun's face completely covered. The places nearby will only see a proportion of the Sun's disk obscured. When the Sun is completely blocked, we call it a total eclipse, when it is partially blocked, a partial eclipse.

A special case occurs sometimes. Neither the Sun nor the Moon are a fixed distance from the Earth; during the course of the month or year, they get closer and further away, and this distance affects slightly the size they both appear in our skies. If an eclipse occurs when the Moon is a long way from the Earth, the tip of the umbra will not reach us and the Sun will not be completely covered. If you are in the right place, a ring of light can be seen around the edge of the Moon: this is an annular eclipse.

Every year there are eclipses, both solar and lunar. The minimum number is four, except in some very rare cases when three solar and no lunar eclipses occur (1908 was the last time that happened) and the maximum number is seven. An eclipse of the Moon is visible from anywhere on the planet that is in night while the eclipse is happening. About three quarters of the planet can see some part of the eclipse, and about a quarter of the planet sees the whole show. So on average you can expect to see one lunar eclipse each year.

Because of the small part of the Earth that is touched by the Moon's umbra during a solar eclipse, the average interval between two total eclipses at the one spot is four hundred years. A notable exception were the 1983 and 1984 solar eclipses, both of which were visible from the region around Port Moresby in New Guinea.

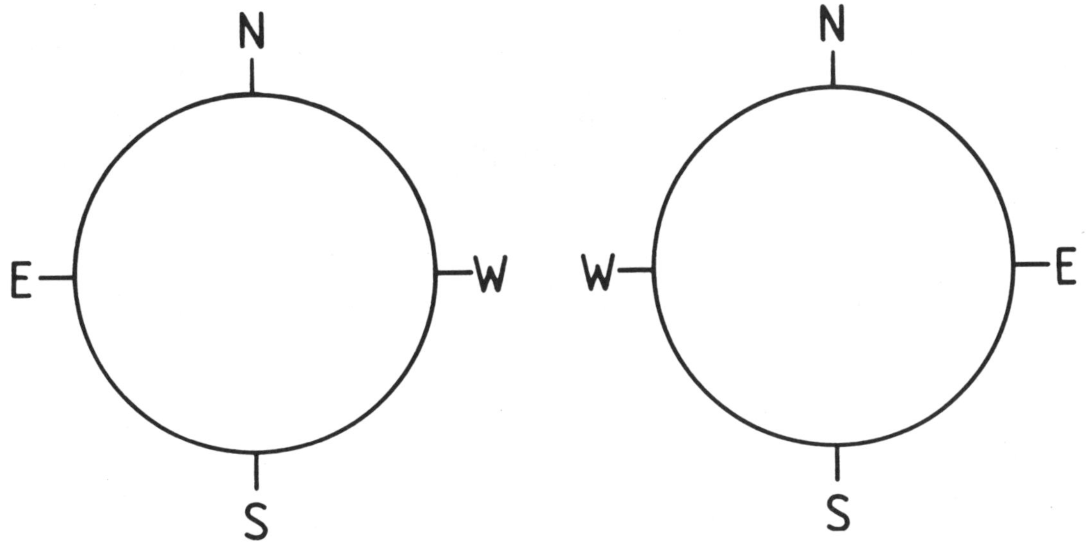

62 This diagram demonstrates the cardinal compass points observable on the Sun using a telescope with an even or odd number of reflecting surfaces. If your telescope has either zero or an even number of reflecting surfaces, use the diagram on the left to find north and south. If your telescope has an odd number of reflections in it, eg a refractor with a star diagonal, use the right hand diagram.

Normally you have to travel to see a total eclipse, and each time there is a favourable one, amateur and professional astronomers travel to the path of totality to watch. In recent years it has become normal for a number of cruise ships to sail into the path of totality also, and many astronomers take advantage of this service.

If you are fortunate enough to view a total eclipse there are a number of things that will happen. As you enter the Moon's penumbra, you will start to see the Moon's disk intrude upon the Sun. As you enter the penumbra further, more and more of the Sun's surface is hidden. Just before it is gone completely, you will see the very edge of the Sun shining between the mountains and valleys of the Moon. This appears as a string of bright pinpoints of light known as Bailey's Beads. They last for a few seconds and then totality is complete.

During the time that the Sun is totally eclipsed it is safe to view it with the naked eye. The Moon blocks all the harmful light rays, leaving you free to view the show. If you are watching a total eclipse, it is important to turn away again as soon as totality ends to ensure the safety of your eyesight.

As totality happens, two solar features not normally visible from Earth can be seen—the first is the Sun's corona, the bright outer atmosphere of the Sun. This appears white, with lots of structure in it; it is often dented near the Sun's polar region. It is difficult to say exactly what it will look like, as it differs from one eclipse to another. The second feature you will see just after totality starts and just before it ends, are the solar prominences. These are huge columns of gas that stream up from the surface of the Sun along its magnetic field lines. They appear pinkish in colour and are often seen best through a small telescope or telephoto camera lens.

As totality ends, the Bailey's Beads again appear, then as you move back into the penumbra the Sun becomes partial, then whole as you move out of the penumbra. The whole event can take two to three hours, but the longest totality can last is eight minutes, and often it is much less than this. The eclipse of 29 March 1987 had a period of totality of less than one second, and even this was seen from only a very small part of the Earth.

Even if you do not travel to see a total eclipse, it is interesting to watch a partial one, but be careful. With only part of the Sun covered, the amount of light coming from it is still enough to cause permanent blindness if you look at it with any form of optical aid, or stare at it. Don't take the risk. Use the eyepiece projection method described above and you will continue being able to observe for many years to come.

The Moon is not the only thing that can pass across the face of the Sun; the two planets between us and the Sun can also obscure our view. Because they are so much further from us neither planet will cause an eclipse. Instead what we see is a small round disk moving slowly across the Sun. Such an event is called a transit. Because of the tilt of their orbits, we again only see these passages when they occur near a node in the orbit. For Mercury the average time between visible transits is seven years, these occurring only in May or November. For Venus the situation is different. Transits come in twos separated by just eight years, but the pairs are separated by over a hundred years.

It was a transit of Venus that brought Captain James Cook and an astronomer to the island of Tahiti in 1769. On the way back to England they discovered New Zealand and the east coast of Australia to top off their work. For those of you wishing to plan a long way in advance, table 8.2 gives a list of the upcoming transits of both Mercury and Venus.

Table 8.2 Coming transits of Mercury and Venus

Mercury	Venus
6 November 1993	7 June 2004
15 November 1999	5 June 2012
	10 December 2117
	8 December 2125

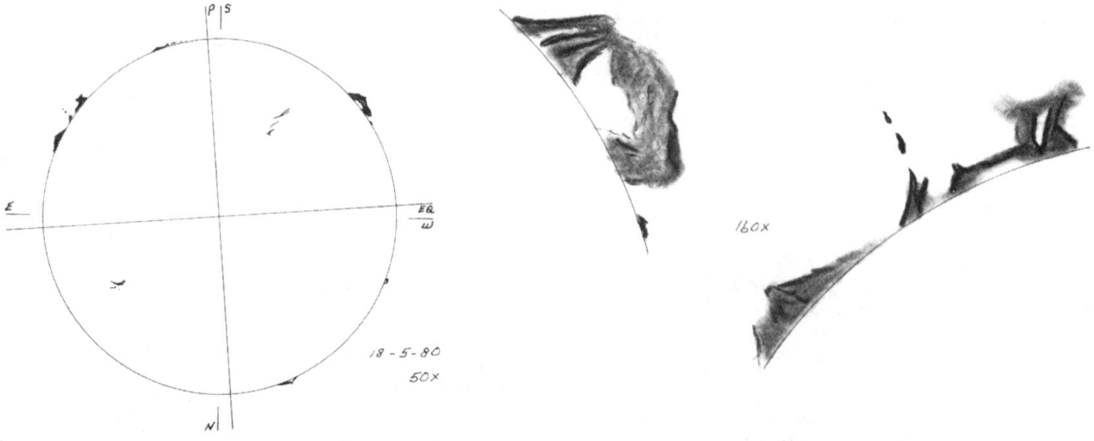

63 Observations of the Sun made using a hydrogen-α filter on 18 May 1980. On the edge of the Sun six features can be seen. More detailed drawings of two of the features, both prominences, show some of the structure visible using a 200mm telescope.

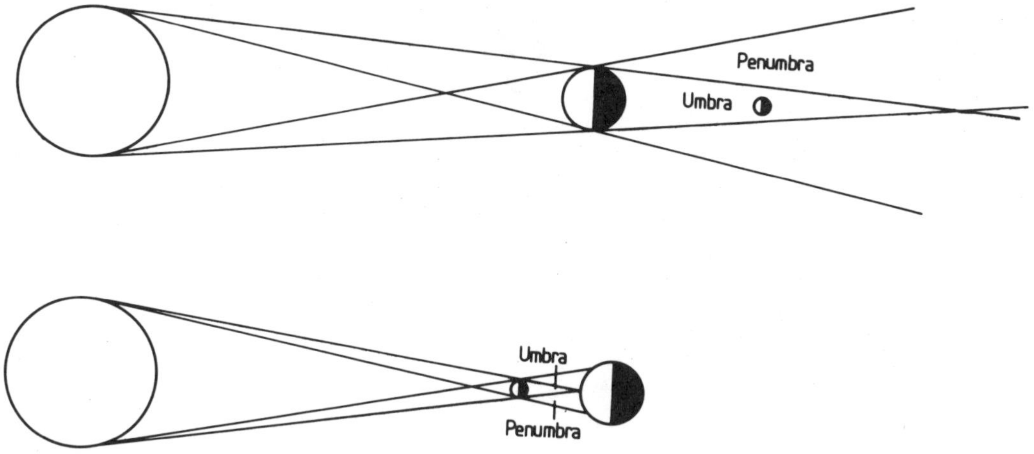

64 The top diagram shows the arrangement of the Sun, Earth and Moon during a lunar eclipse. The Moon is fully within the Earth's umbra, so it is totally eclipsed. The bottom picture shows a solar eclipse. It is easy to see that the Moon's umbra only just reaches the Earth. Totality would be visible only from within this area, while viewers anywhere within the penumbra would see a partial eclipse.

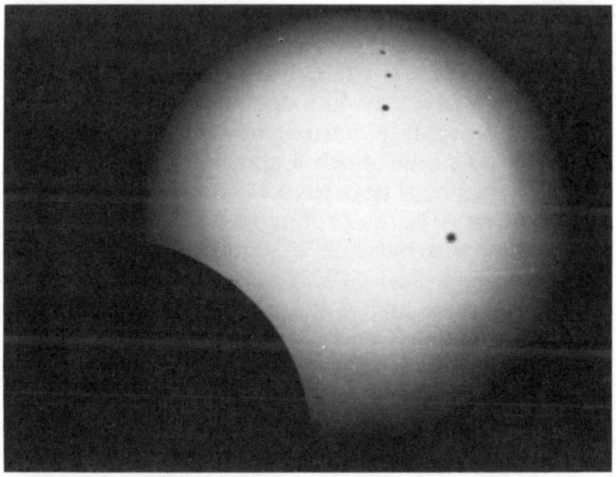

65 This photo of the annular eclipse of 5 February 1981 shows the Moon partially obscuring the disc of the Sun. This 1/60 second exposure taken with an 200mm telescope also shows a number of sunspot groups.
PHOTO: TERRY CLARKE

9

The stars

'Astronomy' means the study of the stars and just as atoms make up all the matter we can see, so stars form all the large-scale structures of the Universe. Our Sun is, of course, a star—it appears so different from all the others because it is so close. The Sun is a fairly average star: there are stars bigger than it and smaller than it, brighter and fainter, yet regardless of their form, all stars run on the same fundamental energy source—nuclear fusion.

Nuclear fusion is an energy source that we use on Earth, not in nuclear reactors (that's nuclear fission) but in the hydrogen bomb. Under the right conditions four hydrogen atoms can be forced together to form one helium atom. This is the basis of fusion. The conditions needed for this reaction are extreme heat and pressure, so extreme that the centre of a star or a nuclear blast are the only places they exist. The joining of these atoms gives off a small amount of energy. Our Sun uses up four million tonnes of hydrogen every second to shine the way it does, yet despite the enormous amount of fuel used, it will remain shining for another 10 000 million years or so.

Scientists think that most stars start their lives in about the same way. Throughout our region of space are large clouds of hydrogen gas, the nebulae. These clouds are generally dark and cold; we only see them because they sometimes affect the light of objects behind them, as we'll discover in chapter 10. These clouds of gas are fairly stable, but sometimes a few of the atoms in them will come together. These atoms may stay together and attract some more, forming a small clump. By gravitational attraction this clump will pull in more material, forming a bigger clump. This process will continue with more and more material forming a bigger and bigger mass of gas. This additional material will cause the pressure and the temperature inside the gas to rise. Eventually it will reach the point where nuclear reactions can start. These begin and a new star is born.

It is the amount of matter that the star begins with that determines how it will spend its life. An average sized star, like our Sun, will settle down into a quiet state, burning its fuel at a steady rate for a long time. Such a star will have a surface temperature of perhaps 6000K and will be yellow in colour. The K, pronounced 'degrees Kelvin', stands for a measure of temperature known as the Kelvin Scale. It is similar to the familiar Celsius scale and is equal to the temperature in Celsius plus 273. It is the scale usually used in astronomy.

A star that is much larger than the Sun will behave differently. It will burn up its fuel at a much faster rate. Instead of lasting ten to twenty thousand million years like the Sun, it may burn for only a few hundred million before it has exhausted its fuel. Such a star will have a surface temperature of about 20 000K and will be blue-white in colour. Rigel in the constellation of Orion is such a star, as are many others in the heavens.

A star which is smaller than the Sun will try to conserve the little fuel it has, burning it slowly over a long period. These stars have very low temperatures at their surfaces, only 3000K. They are red in colour and are generally quite faint.

66 The Rosette nebula in Monoceros is a star formation region about 2600 light years from Earth. Within the nebula, bright stars and dark globules mark the areas where stars are coming into being. PHOTO: TERRY CLARKE

There are many examples of this type of star in the Sun's neighbourhood; they are perhaps the most common type of star.

Eventually all the hydrogen fuel in a star will be exhausted and the star will begin to die. Exactly what happens next will again depend upon how much material the star had to start with. When it begins to run low on hydrogen, a star will begin to use as a fuel the helium it has formed. This makes the star expand to an enormous size. As it expands the surface cools and turns red in colour. These huge stars many millions of kilometres in diameter are called 'red giants'. When our Sun turns into a red giant in the distant future, it will expand out to the orbit of Mars, destroying all the inner planets of the solar system, including anything left on Earth.

Once it reaches the red giant stage, a star will use anything it can to keep going—it will turn helium into carbon and oxygen, and use many other exotic nuclear reactions. Eventually though, there will be no more to burn. With nothing to support its huge size, the star will collapse under gravity until it forms a small dense ball of gas. This will be hot from the left over heat, and will appear white in colour, although rather dim due to its small size. The star has become a white dwarf. From here on the star will slowly cool to a red dwarf and then to a cold dark cinder leaving nothing to see.

The end as a fading red dwarf is not for the big stars. Just as they used their fuel quickly when shining, a large star will, quite literally, go out with a bang. When a star two or more times the size of the Sun runs out of fuel, its core collapses. This causes new nuclear reactions to happen and the star explodes, blowing most of its outer layers off into space. A star that does this is called a supernova and for a time it will shine brighter than a million stars. When such an event occurs near to us, the star becomes so bright that it is visible in the daytime. Unfortunately supernova explosions are rare, the last one in our galaxy occurring in Ophiucus in 1604.

When a star suffers a supernova explosion there are two parts left behind. The first is the sphere of gas thrown off from the surface. This sphere is known as a planetary nebula; it appears as a ring of gas around the star. We shall see more of these in chpater 10. The other remnant is the core of the star itself. Exactly what happens to this will depend upon its weight. If the star were about three times the size of the Sun, the core would turn into a neutron star. This is an object about ten kilometres in diameter which is made from a strange substance. Instead of being formed from normal atoms, a neutron star is made from neutrons, one of the types of particles that make up atoms. Such a star has some weird properties, as we will see in chapter 12.

A star that is much bigger than three times the Sun's size, will not form a neutron star. Because this star will have a huge gravitational pull there is nothing we know of that will stop it collapsing. Eventually it will reach a stage where gravity is so strong not even light can escape from its pull. This has become a black hole. Black holes are some of the most widely sought objects in astronomy; astronomers use them to explain many different phenomena. We will learn more about them also in chapter 12.

At our distance from them we can see nothing of the sizes of stars; they appear as points of light in even the largest telescopes. The only information we have comes from the light they shine towards us. Astronomers can use this light to find

Table 9.1 Some stars, their colours and temperatures

colour	temperature	star
Red	3 000K	α Orionis, α Scorpii, γ Crucis
Orange	4 000K	α Bootes, γ Leonis, α Arietis
Yellow	6 000K	The Sun, ε Geminorum, ß Hydri
Yellowish	7 500K	α Carinae, α Canis Minoris, θ Scorpii
White	11 000K	α Canis Majoris, α Piscis Austrini, α Ophiuchi
Blue-White	25 000K	α Crucis, γ Orionis, ß Centauri

out much about the stars. When light is split into its colours by a prism, astronomers can tell what elements are present where the light was made. Spectroscopy, as this technique is called, has told us that the Sun and stars are all made from hydrogen gas and that helium and other elements are also present. This has lead scientists to develop the models of stars that we accept today.

The amateur astronomer does not have access to the expensive equipment needed to do spectroscopy, but the stars can still offer a wonderful field for study even in the smallest telescope. The light we receive from a star tells us two things; the colour of the star and its brightness. We can easily see stars with colours ranging from blue-white to deep red. These colours can tell us directly how hot the star is. You may be aware of what happens when a piece of steel is put into a fire. As it heats up it begins to glow a deep red, then orange, yellow and finally, if the metal hasn't melted, it will glow white hot. The stars show similar characteristics. Table 9.1 shows the relationship between temperature and colour and some stars which show these colours.

It is difficult to see the colour of a single star because it is so faint. It is by comparing the colour of one star with that of another that the colours really stand out. The best way to compare colours is to find two stars which are close together. Luckily for astronomers there are thousands of examples of this. With the huge number of stars in the sky, it is hardly surprising that some of them will appear close to each other. These are called visual binary stars, visual because we 'see' them as being related when in fact they are not related at all.

As well as the visual binaries, there are some stars which really are related, the true binary stars. We have seen that stars form from clouds of hydrogen gas. Often, more than one star will form in the one cloud. These stars will stay associated with each other throughout their lives. Bound together by gravitational forces, the stars will orbit about each other in the same way as the Earth orbits the Sun. These stars offer a wealth of observational opportunities for amateur astronomers in measuring and plotting the stars' orbits.

Double stars are fascinating objects to view. Often, though the two stars formed in the same cloud, the stars themselves will be quite different. For example, a red dwarf star may be seen with a white companion, a yellow star may have a blue companion. The combinations are enormous and because there are so many and they come in so many shapes and sizes, double stars can be viewed through telescopes of any size. In chapter 14 we have included some of the more interesting double stars to whet your appetite.

Apart from simply admiring the colour contrasts there are a number of observations that amateurs can make on double stars to further our knowledge. The most important of these is the calculation of the period and shape of the stars' orbits. From this information an astronomer can

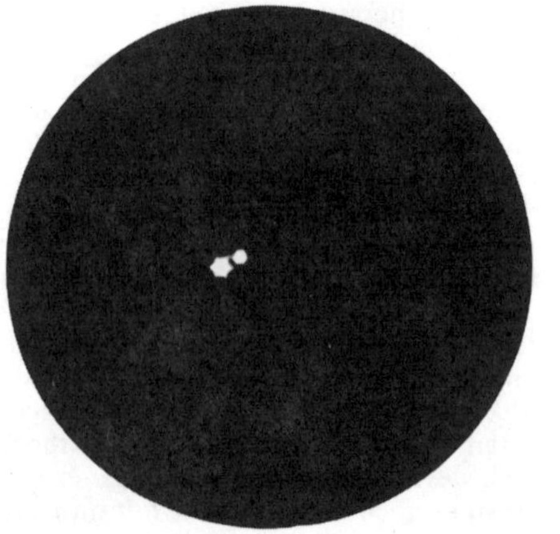

67 This photograph shows the two components of the double star α Centauri. Because of special methods used to separate the components of the star, the primary appears to have six points. PHOTO: KEN WALLACE

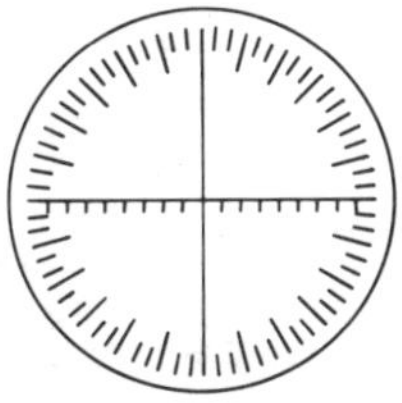

68 This diagram shows the arrangement of lines on a reticle suitable for use in observing and measuring double stars.

calculate the weights of the two stars, the only way we currently have of getting such information.

When observing doubles there are two measurements you need to make. The first is separation, the distance between the two stars. This is measured in seconds of arc. The second is the position angle. That is the angle that a line joining the two stars makes with north. If you want to make these observations you will need to get a special eyepiece for your telescope. Two sorts are available, the micrometer eyepiece and the reticle eyepiece. A micrometer is a delicate and expensive piece of equipment which really needs a large telescope to be used well. The reticle eyepiece is inexpensive and easy to use. I will describe the method of using it to obtain measurements.

The reticle eyepiece is similar in design to a finder scope eyepiece, but where the finder eyepiece has crosshairs, the reticle eyepiece has a network of fine lines and angles. Figure 68 shows the markings on one type of reticle eyepiece. Though there are many other designs, they basically have the same features: a scale to measure distances and another scale to measure angles. Using such a reticle it is possible to measure the separation and position angle of the two stars fairly easily.

Measuring with the reticle is quite simple—it will take you only a few minutes to become familiar with its use, but before you begin you will need to calibrate the separation scale on it. To do this you will need to observe the star μ Crucis, this is a visual double star with a separation of 34.9 seconds of arc and is included as object 59 in chapter 14. Set your reticle so that both the stars in the binary lie along the line with the scale on it. Count the number of divisions between the two stars and record this. When you have finished your observations you can do the calculation to calibrate this scale. Divide 34.9 seconds of arc by the number of divisions you counted: this gives you the number we will call C. Now write the following formula:

$$\text{Separation} = \text{divisions} \times C$$

Replace C with the number you have just calculated.

For an example, if I measure the separation of μ Crucis to be 7 divisions of the reticle then I would make the following calculation. C = 34.9 seconds / 7 = 4.99. So the formula that results is: Separation = divisions × 4.99.

The first problem when observing is to find the double star that you want to observe. Finding the planets is fairly easy, as there are maps drawn to show you where to look, but there are so many double stars that maps are available for very few. Normally double stars are listed simply by their right ascension and declination. Finding an object from only its RA and Dec. is perhaps the hardest part of all astronomy. If you have an equatorial mount which is properly aligned and you have set the setting circles correctly, then you can turn the telescope towards the object simply by setting its coordinates on the telescope's axes. This is fine and is the way that professional astronomers do their work, but for those without equatorial mounts there is another way.

To successfully find objects in the sky you will need a star atlas. The type to buy will depend upon how much observing you intend to do. If you will be observing a lot, and looking for faint objects, then you will need an atlas that shows stars and objects down to the limiting magnitude of your telescope. The bibliography at the end of this book lists some of the atlases available. If you can't justify the expense of a really good atlas, then a simple star map will be sufficient, but will make the job harder.

To begin with you need to work out whether or not the binary you want to observe is in the sky. For the best observations you will need the stars to have an RA within a few hours of local sidereal time. Table 9.2 gives a rough guide to the sidereal time throughout the year, so you can estimate the visibility of the stars. If the stars are going to be visible, use the right ascension and declination to plot the position of the stars on your star chart with a pencil.

In practice, the use of RA and Dec. is much less complicated than the theory behind them that we saw in chapter 5. To use them simply

Table 9.2 Sidereal time compared with universal time

Date		Time												
		18:00	19:00	20:00	21:00	22:00	23:00	00:00	01:00	02:00	03:00	04:00	05:00	06:00
January	5	0:00	1:00	2:00	3:00	4:00	5:00	6:00	7:00	8:00	9:00	10:00	11:00	12:00
	13	0:30	1:30	2:30	3:30	4:30	5:30	6:30	7:30	8:30	9:30	10:30	11:30	12:30
	21	1:00	2:00	3:00	4:00	5:00	6:00	7:00	8:00	9:00	10:00	11:00	12:00	13:00
	28	1:30	2:30	3:30	4:30	5:30	6:30	7:30	8:30	9:30	10:30	11:30	12:30	13:30
February	5	2:00	3:00	4:00	5:00	6:00	7:00	8:00	9:00	10:00	11:00	12:00	13:00	14:00
	13	2:30	3:30	4:30	5:30	6:30	7:30	8:30	9:30	10:30	11:30	12:30	13:30	14:30
	20	3:00	4:00	5:00	6:00	7:00	8:00	9:00	10:00	11:00	12:00	13:00	14:00	15:00
	28	3:30	4:30	5:30	6:30	7:30	8:30	9:30	10:30	11:30	12:30	13:30	14:30	15:30
March	7	4:00	5:00	6:00	7:00	8:00	9:00	10:00	11:00	12:00	13:00	14:00	15:00	16:00
	15	4:30	5:30	6:30	7:30	8:30	9:30	10:30	11:30	12:30	13:30	14:30	15:30	16:30
	22	5:00	6:00	7:00	8:00	9:00	10:00	11:00	12:00	13:00	14:00	15:00	16:00	17:00
	29	5:30	6:30	7:30	8:30	9:30	10:30	11:30	12:30	13:30	14:30	15:30	16:30	17:30
April	6	6:00	7:00	8:00	9:00	10:00	11:00	12:00	13:00	14:00	15:00	16:00	17:00	18:00
	14	6:30	7:30	8:30	9:30	10:30	11:30	12:30	13:30	14:30	15:30	16:30	17:30	18:30
	22	7:00	8:00	9:00	10:00	11:00	12:00	13:00	14:00	15:00	16:00	17:00	18:00	19:00
	29	7:30	8:30	9:30	10:30	11:30	12:30	13:30	14:30	15:30	16:30	17:30	18:30	19:30
May	7	8:00	9:00	10:00	11:00	12:00	13:00	14:00	15:00	16:00	17:00	18:00	19:00	20:00
	15	8:30	9:30	10:30	11:30	12:30	13:30	14:30	15:30	16:30	17:30	18:30	19:30	20:30
	22	9:00	10:00	11:00	12:00	13:00	14:00	15:00	16:00	17:00	18:00	19:00	20:00	21:00
	30	9:30	10:30	11:30	12:30	13:30	14:30	15:30	16:30	17:30	18:30	19:30	20:30	21:30
June	6	10:00	11:00	12:00	13:00	14:00	15:00	16:00	17:00	18:00	19:00	20:00	21:00	22:00
	14	10:30	11:30	12:30	13:30	14:30	15:30	16:30	17:30	18:30	19:30	20:30	21:30	22:30
	22	11:00	12:00	13:00	14:00	15:00	16:00	17:00	18:00	19:00	20:00	21:00	22:00	23:00
	29	11:30	12:30	13:30	14:30	15:30	16:30	17:30	18:30	19:30	20:30	21:30	22:30	23:30
July	7	12:00	13:00	14:00	15:00	16:00	17:00	18:00	19:00	20:00	21:00	22:00	23:00	0:00
	15	12:30	13:30	14:30	15:30	16:30	17:30	18:30	19:30	20:30	21:30	22:30	23:30	0:30
	22	13:00	14:00	15:00	16:00	17:00	18:00	19:00	20:00	21:00	22:00	23:00	0:00	1:00
	30	13:30	14:30	15:30	16:30	17:30	18:30	19:30	20:30	21:30	22:30	23:30	0:30	1:30
August	6	14:00	15:00	16:00	17:00	18:00	19:00	20:00	21:00	22:00	23:00	0:00	1:00	2:00
	14	14:30	15:30	16:30	17:30	18:30	19:30	20:30	21:30	22:30	23:30	0:30	1:30	2:30
	22	15:00	16:00	17:00	18:00	19:00	20:00	21:00	22:00	23:00	0:00	1:00	2:00	3:00
	29	15:30	16:30	17:30	18:30	19:30	20:30	21:30	22:30	0:30	1:30	2:30	3:30	4:30
September	6	16:00	17:00	18:00	19:00	20:00	21:00	22:00	23:00	0:00	1:00	2:00	3:00	4:00
	13	16:30	17:30	18:30	19:30	20:30	21:30	22:30	23:30	0:30	1:30	2:30	3:30	4:30
	21	17:00	18:00	19:00	20:00	21:00	22:00	23:00	0:00	1:00	2:00	3:00	4:00	5:00
	29	17:30	18:30	19:30	20:30	21:30	22:30	23:30	0:30	1:30	2:30	3:30	4:30	5:30
October	6	18:00	19:00	20:00	21:00	22:00	23:00	0:00	1:00	2:00	3:00	4:00	5:00	6:00
	14	18:30	19:30	20:30	21:30	22:30	23:30	0:30	1:30	2:30	3:30	4:30	5:30	6:30
	21	19:00	20:00	21:00	22:00	23:00	0:00	1:00	2:00	3:00	4:00	5:00	6:00	7:00
	29	19:30	20:30	21:30	22:30	23:30	0:30	1:30	2:30	3:30	4:30	5:30	6:30	7:30
November	6	20:00	21:00	22:00	23:00	24:00	0:00	1:00	2:00	3:00	4:00	5:00	6:00	7:00
	13	20:30	21:30	22:30	23:30	0:30	1:30	2:30	3:30	4:30	5:30	6:30	7:30	8:30
	21	21:00	22:00	23:00	0:00	1:00	2:00	3:00	4:00	5:00	6:00	7:00	8:00	9:00
	28	21:30	22:30	23:30	0:30	1:30	2:30	3:30	4:30	5:30	6:30	7:30	8:30	9:30
December	6	22:00	23:00	0:00	1:00	2:00	3:00	4:00	5:00	6:00	7:00	8:00	9:00	10:00
	14	22:30	23:30	0:30	1:30	2:30	3:30	4:30	5:30	6:30	7:30	8:30	9:30	10:30
	21	23:00	0:00	1:00	2:00	3:00	4:00	5:00	6:00	7:00	8:00	9:00	10:00	11:00
	29	23:30	0:30	1:30	2:30	3:30	4:30	5:30	6:30	7:30	8:30	9:30	10:30	11:30

think of a street directory. If you want to find a certain place, you look up its location in the directory; it may say map 4, position K 12. On map 4 you would look along the line marked K until it crossed the line marked 12. That's the position of the place. It works the same way for stars. An easy double star is α Centauri; its position is given as 14h38m18s −60°45′28″. Find the map that has 14h and −60° on it. On this map find 14 hours RA then move about half way towards 15 hours to 14h38m. Then move down to about −60° and mark the position. If you've got it right, α Centauri should be close by.

When you are ready to observe, take your map

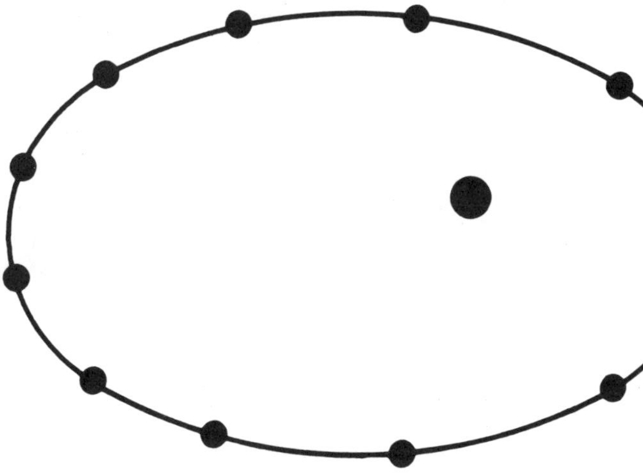

69 This diagram shows the motion of the secondary component of a double star about its primary. It is interesting to note that the positions given for the secondary are all equally spaced in time. One of Kepler's laws of motion says that the secondary star will move fastest when it is near the primary and slowest when it is furthest away. It is possible for amateurs to draw diagrams like this by observing double stars over a long period, then plotting their measurements of separation and position angle.

and red torch outside with your telescope. Using the bright stars on the map as a guide, identify the area in the sky that includes the double star you are looking for. Now is the time that your knowledge of the sky will help you. After you have found the area try to identify the star, if it is bright enough. If it is not find a nearby star that is bright enough to be seen. Use the technique given in chapter 6 to point your telescope at the star you have identified. If you now have the double in the field of view, you're finished. If you have pointed the telescope at a nearby star you will need to do a little more.

Once you are near to the double, or any other object you want to see, you need to start 'star hopping'. This is the name that amateurs give to moving from a bright star starting point to a faint object. To use this technique it is important to use a low power eyepiece. With the bright star in the middle of the field of view, work out which way the field of view is pointing. To do this you will need to identify some of the faint stars in the field with those on your map. Then work out which way you will need to move to get to the object. Slowly move the telescope in the required direction until the object appears in the field. Centre it and you're ready to observe. Now it is time to switch to a higher observing power, or to your reticle eyepiece. After you have had some experience using this technique you will be able to star hop in two or three steps using faint stars as guides to locate some really difficult objects.

One complication that can happen with this simple system occurs if your telescope has an odd number of reflecting surfaces. A refractor with a star diagonal is such a combination, but a normal Newtonian with just the primary and secondary mirrors is not. An odd number of reflecting surfaces will cause the field of view in the telescope to be reversed with respect to your charts. If this is a problem you can copy your map onto tracing paper and then use the copy from behind.

With the reticle eyepiece in position you can start measuring. When looking at a pair of stars we designate one of them as the primary star and the other as the companion. The brighter of the two stars will be the primary, but if the two are of equal magnitude then you can choose either. Using the reticle in the eyepiece place the primary star at the centre by carefully adjusting the telescope's controls. Turn the eyepiece so that both stars are on the main axis. Then measure the separation of the two stars with the scale on the reticle.

Once you have observed the distance shown by the reticle, write it down, along with the usual details of universal time and date. When you have the separation recorded, keep the telescope still and let the stars drift through the field of view. As they move, the primary star will drift in a certain direction; you will be able to read this from the angle scale on the reticle. Record this reading with the other details of the observation.

When you have finished your observations for the evening you can turn your measurements into a record of angular separation and position angle. Use the formula you wrote down when

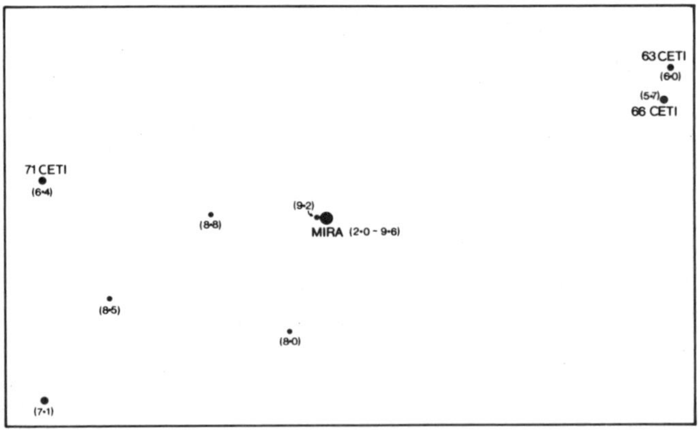

70 A comparison chart for the famous variable star Mira, object 4.

you did the calibration to turn the reticle reading into a separation. Normally you only record separations to one decimal place. The position angle comes from adding 90° to the reticle reading. If this results in an angle of more than 359° subtract 360° from your calculation. Don't try to do the calculations when you're observing, it's too easy to make a mistake. Wait until the next day if you want, but don't leave them much longer.

When you have a collection of records you can pass them on to one of the many organisations involved in the reduction of observations into scientific results. Your local astronomical society can put you in touch with one of these groups. They may well have a programme of observations of their own running to which you can contribute.

Another type of star that is of interest to the amateur astronomer is the variable. We have seen that most stars are stable for the greater part of their lives, but some stars are not—they vary in brightness. The reasons for this are numerous; some stars expand and contract, others get hotter or colder—these are intrinsic variables. Some other stars are in reality very close binary stars. They are so close that no telescope will show two separate stars; these are extrinsic variables. Regardless of the exact mechanism that causes it, these stars appear to change their brightness. Both professional and amateur astronomers spend hours observing these objects to learn their secrets.

The variable stars divide into a number of different classes, these are listed in appendix 3, but broadly there are two observational types, those that have regular periods and those with irregular variations. Figure 71 shows the graphs that result when the brightness of a star is recorded over a long period of time. As you can see, the results are quite varied. It is these graphs that professional astronomers use to help them formulate theories about what is happening in these stars. Amateur astronomers around the world play a part in helping professionals to collect these observations.

The equipment needed to measure the magnitude of a star directly is quite expensive and needs a large telescope to be used. To remove the need for such equipment, amateur astronomers use another method known as magnitude estimation. It is difficult to accurately guess the magnitude of a star just by looking, so amateurs cheat. They compare the magnitude of a variable star with those of the fixed stars near it. If you decide to take up variable star observing you will need some finder charts for them. These charts are available from the groups that organise world-wide variable star observations; you can probably get some through your local society.

A finder chart, such as that shown in figure 70, has the variable star marked, as well as stars around it of similar magnitudes. To use these charts involves a number of steps. Firstly you must find the star—use the same method as that given above for double stars. Use the lowest power eyepiece you have as this makes the object easier to find; it will also allow you to get all the comparison stars in the same field. Once you have found the field, identify the variable star and the comparison stars. Remember that the field will be reversed if you have an odd number of reflections in your telescope.

Once you have all the stars identified you can

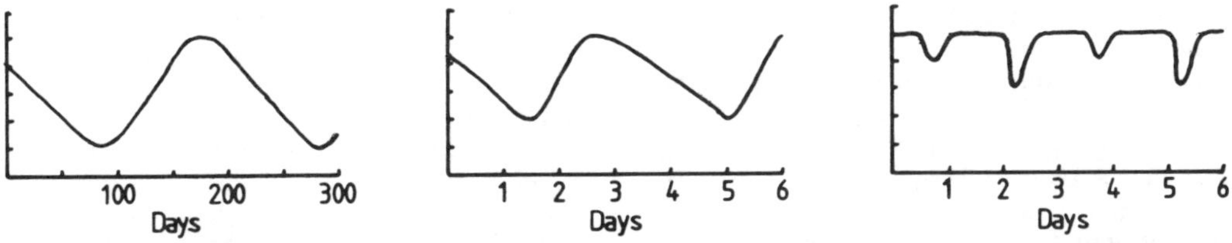

71 Here we see light curves for three different types of variable stars. The first is for a long period variable, the second for the famous Cepheid type variable and the third, the distinctive shape of an eclipsing binary. Amateur astronomers throughout the world observe variable stars over many years, producing graphs such as these and increasing our knowledge of the stars.

begin estimating by finding the comparison star that is just brighter than the variable. Next find the comparison star that is just fainter than the variable. Write down the magnitudes of these two stars. That's half the job done. If you are lucky, the variable star may be exactly the same brightness as one of the comparison stars; if this is the case just record this fact on your observation. If it doesn't match a comparison star then you will need to do an estimate.

When estimating, the idea is to get as close to the variable's magnitude as you can. To do this you divide the interval of brightness into steps and then place the variable on one of these steps. Steps the size of a quarter of the difference in magnitude are the easiest to use. Either the variable will be half way between the two comparison stars in brightness, or it will be closer to one than the other. Mark which of the stars you think the variable is closest to, equal to, or half way between, on your observation along with the usual data.

When you have finished observing, you can turn your observations into an estimate of the variable star's magnitude. Use the formulae given in appendix 4 to do this. Variable star observing is a part of the hobby that takes a lot of dedication and effort, yet it is one that can be of real scientific value. The methods used when you become experienced at variable observing become much more precise and difficult to master. Practice at using the simpler methods will stand you in good stead for further observations. Once you have become proficient at variable star observing, you may like to contact one of the international groups. They are always after dedicated amateurs to take part in their programs.

One further contribution that amateurs can make to professional astronomy lies in the discovery of novae and supernovae. Supernovae are such brilliant objects that, were one to occur near to us, there would be no discoverer as it would be so obvious. Novae are, however, a different matter. These occur unexpectedly, and though they increase immensely in brightness, it may not be enough to be immediately obvious. A good knowledge of the sky is definitely an advantage in nova searching. As for comets, you must be always on the lookout for a new object in the sky. If you find such a new object, and you're sure that it is not a planet or a mistake on your part, contact your local astronomical society or professional observatory immediately: you may have found a nova.

Each year many stars go nova, but most of these are too faint to be seen except with large telescopes. An organisation in the United States runs a service to inform astronomers of new discoveries of novae and other astronomical phenomena. Many astronomical societies receive this service and pass the information on to their members to keep them up to date.

Before leaving the subject of stars, it is important to note one of the major contributions they have made to our understanding of the universe. Stars have let us measure the size of our universe. If you consider the obviously vast distances involved in astronomy, you will probably wonder how they were first measured. After all, if you can't get to a place, how do you measure its distance?

The distance to the stars has always intrigued astronomers, but it wasn't until fairly recently that any accurate measurements were made. Although William Herschel suggested a method of measuring the distances to stars, it was the German astronomer and mathematician Friedrich Bessel who, in 1837, did the first successful observations. Bessel used a system of measuring distances called the parallax method. The method itself was not new—it had been used by surveyors for centuries—but Bessel was the first to make observations accurate enough for it to be used in astronomy.

You use the parallax method of measuring distances every day, but are probably unaware of it. Try a simple experiment. Hold your hand out in front of your face with your index finger up. Just by looking you know that your finger is closer to you than the objects behind it. Now close your left eye. Note the position of your finger with regard to the objects in the distance. Now close your right eye and open your left. Now note the position of your finger; it appears to have moved to the right. This is caused by your eyes being a certain distance apart. It's called the parallax effect.

If you know the distance between your eyes and the exact angle throught which your finger appears to move, it is a fairly simple matter to calculate the distance to your finger. Similarly you can calculate the distance to any other object just by having two eyes. Our brains do this automatically, making us able to see depth. Because the distance between our eyes is fairly small, parallax only works to a certain distance. For example, it is difficult to tell which of two distant buildings is closer if they both appear of different sizes. Surveyors, when they are using this method, increase the distance between the two points from which they observe. This increases the angle that the object seems to make, and hence lets them use this method to much greater distances.

It was this parallax method that Bessel adapted to find the distance to the stars. To begin with he made the assumption that some stars were much closer than others, and that these were the ones that he would try to measure. His first problem was to find a base line long enough for him to

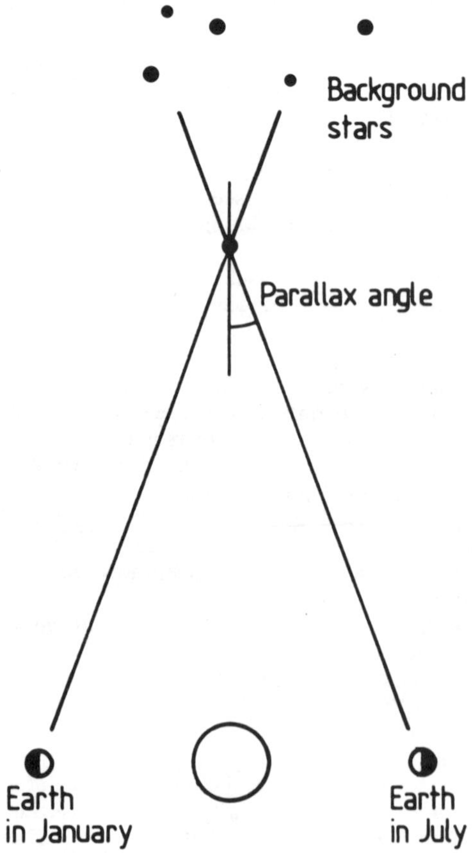

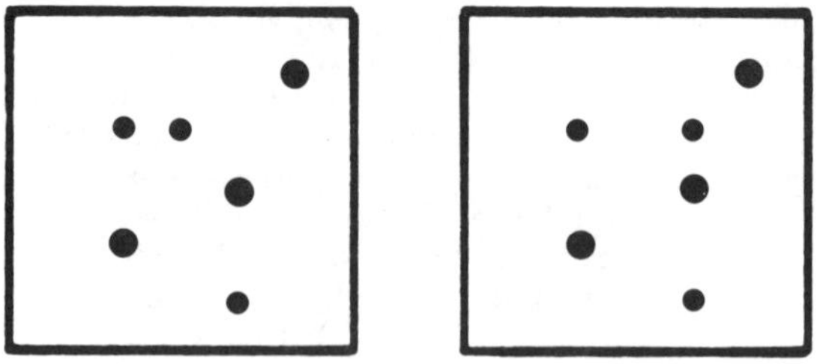

72 This diagram shows the method of parallax used by astronomers to measure the distance of nearby stars. Using the orbit of the Earth as a baseline, observations of the positions of stars are made at intervals of six months. Then by comparing the observations (see above) the distance to a star can be calculated.

make the measurements accurately. He knew that he would be dealing with immense distances, so to get the longest base line he decided to use the Earth's orbit. He would measure the positions of the candidate stars once, and then again six months later when the Earth had moved 300 million kilometres.

Bessel chose a number of stars to begin with, mainly bright stars as he thought that these would be closest to the Earth. His first success was with the star 61 Cygni. He measured the angle that this star made with the background stars to be 0.676 seconds of arc, an incredibly small amount. From this measurement, Bessel calculated that the star was six light-years distant. He then found he had made a mistake and adjusted this to eleven light-years in 1838.

Further measurements made by Bessel and others gave astronomers accurate distances to many nearby stars. It was the first time that scientists had had any idea of the size of the universe. The story does not end there, but we need to know something of the other inhabitants of the heavens before we can continue.

10

Deep-sky observing

Stars and planets are not the only objects visible in the night sky. There are also the 'deep-sky' objects, so named because many of them are much further away from us than the stars. These objects tell astronomers about the universe in which we live. For an amateur without the expensive equipment needed for the precise measurements needed for research, the deep-sky objects offer some of the most spectacular views of all.

Deep-sky observing divides into two parts, objects which are inside our galaxy, and objects which are outside it. Our galaxy is a large disk of stars of which the Sun is but one. You can see it with the naked eye as the large fuzzy band of light stretching across the night sky, called the milky way. The deep-sky objects within it are the nebulae and clusters.

We have seen that stars form in clouds of gas throughout the galaxy—these clouds are visible through the telescope, some even with the naked eye. They are the nebulae, from the Latin word for cloud. There are three types of nebulae visible with a telescope: emission nebulae, reflection nebulae and dark nebulae.

An emission nebula is a birthplace of stars. Within it are hot blue and white stars that have just started shining. These stars give off heat and light to the gas surrounding them, causing it also to shine. One of the most famous of the emission nebulae, the Orion nebula, is so big and bright it can be seen with the naked eye. It appears as the middle star in Orion's sword. Within this cloud of glowing gas, a telescope will show four new stars, the trapezium. These stars were born within the nebula in the past few million years.

Using special infra-red telescopes, astronomers have found more baby stars in the nebulae. These protostars as they are called, are not yet hot enough to give off visible light. Astronomers detect them by the heat they give off as they collect gas from the cloud. The Orion nebula is a fascinating object to watch; it is included in chapter 14.

Nebulae do not only mark the birth of stars, they also mark their death. A second type of emission nebula is formed when a star explodes as a nova or supernova. As we saw in chapter 9, these explosions cause the star to lose some of the outer layers of its atmosphere. As these layers expand out into space, the remnant of the star causes them to glow. These types of nebulae often form rings or disks around their stars. Others form disorganised blobs around their central stars.

When early astronomers viewed these objects with their telescopes, they thought they looked a little like planets, so they called them planetary nebulae. The name has remained even though these objects have nothing to do with planets. Perhaps the most famous, certainly the most studied, of these nebulae is the Crab nebula in Taurus. This object is shown in figure 89. It looks like a fairly disorganised collection of gas. What makes the Crab so important is that we know exactly when it was formed.

The Chinese have a long history of astronomical observations—on 4 July 1054 they reported seeing a new star shining forth in Taurus. This

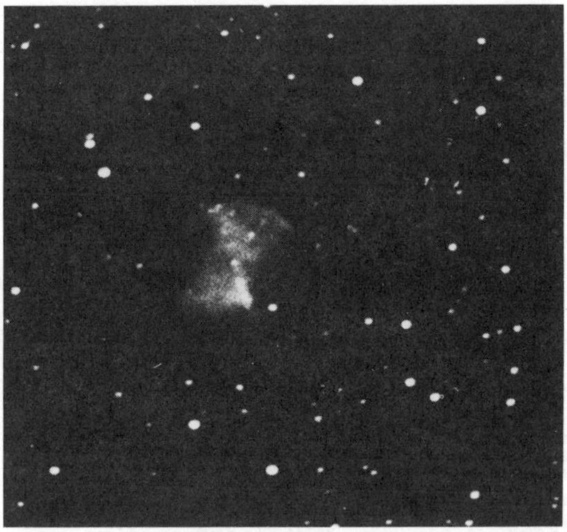

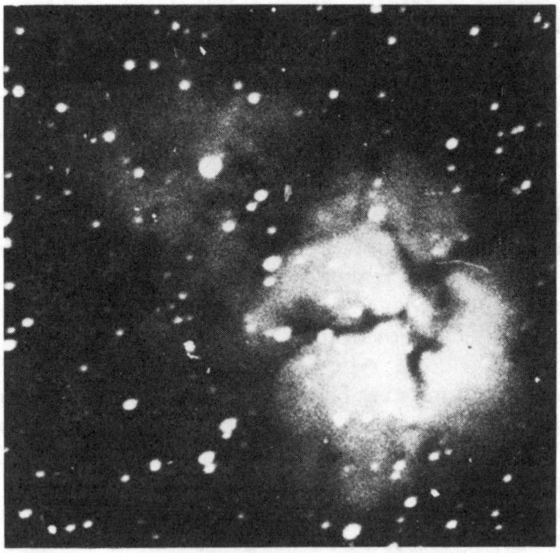

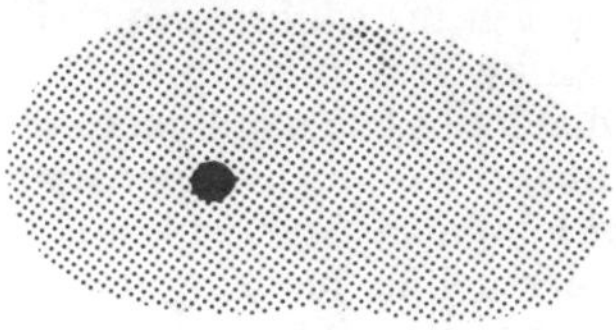

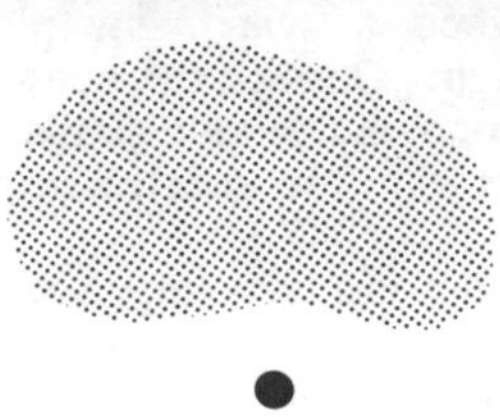

73,74,75 Here we see the three types of nebulae visible in our skies and the way they are formed. Above left is M27, the Dumbbell nebula, an emission nebula made to shine by the hot star within its gas, as shown in the diagram below it. Above right is the Trifid nebula. Although part of it is an emission nebula, the fainter part to the upper left is a reflection nebula reflecting light from nearby stars, as shown below it. Opposite we have a dark nebula obscuring the light of the star clouds behind it as in the last diagram. PHOTOS: 73 AND 75, JOE CAUCHI; 74, TERRY CLARKS

new star was the result of a supernova explosion. The reports say that the star remained visible for a number of months and that at its brightest it could be seen in the middle of the day. The nebula that resulted from this explosion was recorded on 12 September 1758 by the French astronomer Messier. Many photographs have been taken of the Crab and these show that it is still expanding. Pictures taken as little as twenty years apart clearly show the change in size of this nebula. We have of course included the Crab in the collection of objects.

The second type of nebula we see is the reflection nebula. These are clouds of gas that do not have any stars within them, and therefore do not glow by themselves. Often these clouds of gas are near to other bright stars, so instead of glowing themselves, they reflect the light of the nearby stars towards us. It is similar to what happens when a searchlight plays over the clouds at night. Through the telescope, emission and reflection nebulae look similar. Professional astronomers can use spectroscopy to tell the difference.

Spectroscopy is a technique where a prism or a diffraction plate is used to split the light of a star or other object up into its component colours. Every material that gives off light does so in slightly different ways. From looking at the image from the spectroscope, called a spectrum, astronomers can tell what elements were present when the light was emitted. When astronomers look at an emission nebula, they see that the light was given off by very sparse hydrogen gas; when they look at a reflection nebula, they see only the reflected spectrum of a star.

The last of the nebulae are the dark nebulae. Like the reflection nebulae these clouds do not contain any stars to make them glow, but neither do they reflect light. Dark nebulae are just that, dark. We detect their presence by the fact that they block the light of stars and other objects that lie behind them. It is possible that we do not know of many of the dark nebulae, as we don't notice them blocking anything. There are two dark nebulae of special note. One is the Coal Sack. If you look at the Southern Cross on a clear, dark night in the country, you will see the band of the Milky Way running by it. You will also notice what looks like a dull piece of sky just to one side of the cross. This appears to be an area of sky without many stars; it is not. In reality it is a large dark nebula blocking the light of all behind it. The few stars which do appear in the dark region are in front of the cloud. Figure 110 is a photo of this region of sky.

76 This photo of the Horsehead nebula in Orion clearly shows how the nebula got its name. PHOTO: KEN WALLACE

Another famous dark nebula is the Horsehead nebula in Orion. The nebula's name comes from the fact that it looks very much like the horse's head used on the knight pieces in chess. The Horsehead is a dark nebula that intrudes into an otherwise bright emission nebula. From the Earth it appears as if part of the bright nebula is simply missing, but astronomers have measured the distance to the dark cloud as being about 400 light-years closer to the Earth than the bright nebula behind it. Unfortunately the Horsehead is almost invisible in even a large telescope; it takes a long exposure photograph to show up its details, so we did not include it among the objects in this book. Figure 76 is such a photograph. If you are taking guided exposures of the Orion area of the sky you will usually find the Horsehead on the photograph.

When looking at nebulae, both through the telescope and in photographs, it is possible to imagine that these are very dense areas of gas, almost as thick as the rain clouds we see in our skies. In fact the clouds are very sparse. If it were possible to collect some of the material making up one of these nebulae, it would be almost impossible to distinguish it from a vacuum, it is so thin. The only reason that a nebula looks thick is that there is so much of it. An average-sized nebula is tens of light-years in size. We are looking at a huge amount of gas when we observe one. In fact it has been speculated that most of the material in the universe may be tied up in the nebulae rather than the stars.

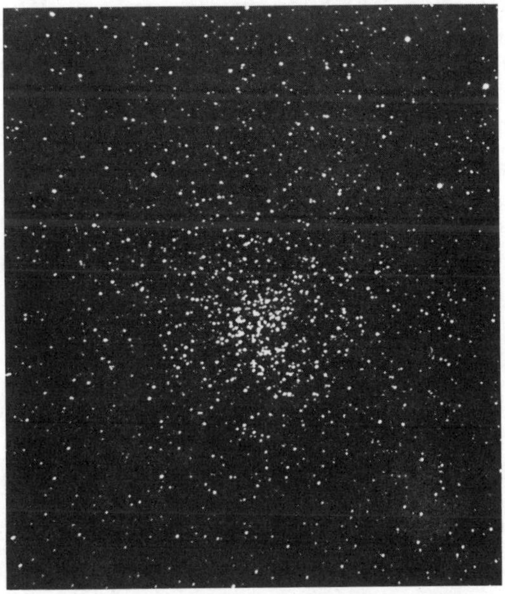

77 M37, an open cluster in Auriga. PHOTO: JOE CAUCHI

We have seen that stars sometimes group together in pairs or other small groups; they do however also form up in larger groups. The smallest of these groups are the open clusters, collections of a few hundred stars which are loosely linked to each other by gravity. Viewing an open cluster is best if you use low power—in this way you will see all the stars in the group, and will better see the stars with small magnitudes.

Open clusters give a good opportunity to compare the colours of stars. One cluster that is good for this is near the star κ Crucis. Because of the large number of coloured stars in this group it is known as the Jewel Box. The Jewel Box contains stars which are purple, red, blue, yellow and white and because of their closeness to each other, the colours are much more noticeable than usual. The Jewel Box is included with the objects at the end of this book.

Another open cluster, probably more famous than the Jewel Box, is the Pleiades. These are a group of stars in the constellation of Taurus. In the group there are seven stars visible to the naked eye, so the group is also known as the seven sisters by some. The Pleiades are a fascinating object in a small telescope, for once you apply some power to them many more stars become visible. There are more than a hundred stars in the Pleiades; a telescope will reveal many of them. Scientists have studied the Pleiades quite extensively and have learned many things from them. They are surrounded by a diffuse emission nebula that is being made to glow by the Pleiades stars. In this nebula astronomers have found yet more stars forming, stars which will eventually add their light to that of the others.

Another grouping of stars that exists is the globular cluster. This is a huge sphere made of hundreds of thousands of stars. Globular clusters are just outside the main mass of local stars, forming a halo around the Milky Way. Astronomers have found hundreds of these objects scattered throughout the sky, but the best by far is visible nearly all year round from the southern hemisphere.

Before telescopes were invented, when astro-

78 M13, a fine globular cluster in Hercules, is regarded by most to be the best of the northern hemisphere's globulars. PHOTO: JOE CAUCHI

nomers were cataloguing and naming the stars and constellations, they named the twenty-fourth brightest object in Centaurus, ω Centauri, as they thought it was a normal fourth magnitude star. When astronomers turned their telescopes towards it in the eighteenth century, they discovered that it was not in fact a star, but a huge ball of stars, thousands of them packed next to each other. ω Centauri is a globular cluster. This cluster of stars is so big it is hard to comprehend. Although it is 17 000 light-years from us, it has a diameter of 30′ in the sky; that's the same size as the full Moon. With the naked eye it appears as a small hazy patch. When Halley's Comet was in our sky, it passed close by ω Centauri and many people had difficulty telling the two objects apart. In fact some of the media reported the appearance of another comet by confusing the two.

In the region about the Sun, the average distance between stars, not including doubles, is about four light-years. Inside a globular cluster the average is less that half a light-year. The sky inside a globular would be filled with bright stars all year round. Through the telescope we are able to resolve some globulars into their component stars. Guided photographs will show you many more.

Globular clusters are important objects in astronomy—it was through using them that we first gained some idea of the shape and size of the Milky Way, our galaxy. It was the American Harlow Shapley who first made a map of the positions of all the globular clusters. He noticed that the globulars seemed to be concentrated in the direction of Sagittarius. He conjectured that if the globulars made a halo around our galaxy, they would tend to cluster near the centre of it because of perspective. By further calculations he was able to show that if this were the case, the centre of our galaxy would be in the direction of Sagittarius, right where the Milky Way appears to be densest. He estimated that our Sun would be 30 000 light-years from this centre. More recent observations using sophisticated techniques have shown that Shapley was correct. Other observations of nearby galaxies have shown that they too are surrounded by families of globulars.

I have mentioned galaxies quite a bit in this chapter, but what exactly are they? Galaxies are the largest single structures known. Made of millions of stars, they form islands of light in the universe. For many thousands of years people have known of the bright band of light that stretches across the night sky. Because of the vague resemblance to a stream of milk it became known as the Milky Way. As man progressed in astronomy, he recognised the Milky Way as a vast collection of hundreds of millions of stars of which the Sun was only one member. We have seen how Shapley, working in 1918, gave us the size and shape of the Milky Way. He and others calculated its shape as a flattened disk with a diameter of 100 000 light-years.

For many years astronomers had been aware of fuzzy patches which looked like nebulae, but were vaguely spiral in shape, and whose spectra did not match those of other emission and reflection nebulae. Some astronomers had speculated that these nebulae were actually distant groupings of stars and this was fairly much accepted. In 1919 it was observed that novae which occurred in these nebulae were about ten times fainter than those in our galaxy. This meant that the spiral nebulae must be about a million light-years away. This was well outside the known bounds of our galaxy. Scientists had suddenly increased our knowledge of the universe's size tenfold.

In 1905 Henrietta Leavitt was observing a special type of variable star called a Cepheid variable. These stars brighten and dim in a set pattern that relates directly to their brightness. Using these stars it is possible to calculate the distance to the star simply by measuring the time

79 The star clouds in the region of Scorpius and Sagittarius are a marvellous sight from a good dark site. The very dense region near the bottom of the photo hides the centre of our own galaxy from view. PHOTO: KEN WALLACE

between variations. Another American, Edwin Hubble, took Ms Leavitt's calculations and began observing Cepheid variables in a large spiral nebula in Andromeda. His observations of the Cepheids lead him to estimate that the Andromeda nebula was in fact a vast flattened disk of stars similar to our own Milky Way and three-quarters of a million light years from us.

Further observations revealed many other nebulae which were in fact more stars. The term 'galaxy' was used to name these objects. Very quickly it was found that some important errors had been made in the initial calculations and that the Andromeda Galaxy was in fact two million light-years distant and much larger than our own galaxy. The Magellanic Clouds of Ms Leavitt were again measured and found to be small companion galaxies to our own. These two small galaxies have been the subject of intensive research from Australian observatories for the past few decades. We now know that many of the assumptions made by Ms Leavitt were in fact wrong, but her results were on the right track and lead to one of the fundamental discoveries in astrophysics.

Over the years many more galaxies have been identified. After the spiral galaxies had been catalogued, it was noticed that some of the round objects which looked like faint globular clusters also contained Cepheid variables. These turned out to be vast elliptical galaxies, many times larger than our own and much older. Many other sizes and shapes of galaxies were discovered, and a naming system grew to help classify them. Firstly the galaxies were split into three main classes, the spirals, the ellipticals and the irregulars. The spirals were then further divided into normal spirals and barred spirals. With a small telescope many of these classifications are academic for the galaxies are too faint to make out any detail at all. Long-exposure eyepiece photography, as we will see in the next chapter, is the only way an amateur can hope to see these distinctions.

The last step, so far, in our exploration of the size of the universe began in 1929 when Edwin Hubble published an amazing theory. Hubble had observed many galaxies and had looked at their spectra. What he found was that the lines made by the elements were not in the correct places; they had moved a little. In nearly all cases the movement was towards the red end of the spectrum. A phenomenon known as the Doppler Effect was the cause. You are probably familiar with the Doppler Effect. When a police car or fire engine races past you, you will have noticed that the pitch of the siren drops. This is caused by the frequency of the sound moving towards the

80, 81 These two photographs show two forms of spiral galaxies. On the left is M61, a face-on spiral in Virgo. On the right is another of the Virgo galaxies, M104. Also known as the Sombrero Hat galaxy, M104 shows a prominent dark dust lane separating the top and bottom halves. These dark lanes are typical of most spiral galaxies. PHOTOS: JOE CAUCHI

low end of the scale as the source of the sound moves away from you. By careful measurement you would be able to measure exactly how fast the police car or engine was going. This is also how speed-detection radar works; the machine measures the increase in the frequency of the beam of microwaves it gives out. This increase tells the machine how fast you are going.

Hubble found that the further a galaxy was from us, the faster it was travelling away, as if all the galaxies were fleeing our own. This doesn't mean that we are in a special position though; it simply means that our universe is expanding. Consider a balloon being blown up. If you were to draw dots on its surface while it was deflated, then blew it up, it would appear as if every dot were moving away from every other dot. A similar thing is happening in our universe; it is getting larger. Nowadays we use Hubble's Law to work out the distance to the furthest objects. By recording their spectra, their distance can be easily found.

The unfortunate thing about the galaxies is that most of them are too faint to be seen with small telescopes, though there are some exceptions. The two Magellanic Clouds make for good viewing even with the naked eye or binoculars because they are so close. Although you won't be able to see many stars in them you can make out some nebulosity. In the Large Magellanic Cloud there is the largest nebula known. Called the Tarantula nebula it is over 20′ of arc wide. The Tarantula shows up very well on wide-angle guided photographs of this region of sky. These photos will also show up areas of nebulosity and other features of the two clouds.

The Andromeda galaxy, the best known of the galaxies, is a disappointing sight from much of the southern hemisphere, as it is so far north in the sky. If you do observe it, you will see a fuzzy disk almost 1° in width with two smaller galaxies nearby. The Andromeda galaxy also has two companions like our Magellanic Clouds. But while Andromeda's are elliptical, ours are a barred spiral and irregular.

One observing program that is open to amateur astronomers and deals with galaxies is the supernova search program. We saw in chapter 9 that some stars end their lives with a big bang, but this is rare on the scale of a human lifetime, the last one in our galaxy happening in 1604. Scientists have estimated that in any galaxy there should be one supernova explosion every two or three hundred years, on average. Therefore, if we were to observe a few hundred galaxies, we should be able to find at least one with a supernova each year. The supernova search program involves amateurs from all over the world observing a range of bright galaxies at regular intervals, looking for supernovae.

Supernovae are so bright when they go off, that for a short time they give out more light than all the rest of the stars in the galaxy combined. Figure 82 shows a photograph of the Large Magellanic Cloud. In the shot is a supernova. It was discovered on 23 February 1987 by the New Zealand amateur astronomer Albert Jones and others. It is the brightest supernova since Kepler's in 1604. It is also the only supernova visible to the naked eye in over a hundred years. The supernova search programme is a valuable contribution to astronomy and is recognised as

82 At lower centre, this photograph of the Large Magellanic Cloud shows the supernova which became visible in February 1987. The light, which first arrived on 23 February, took over 170 000 years to reach us from the LMC. The photograph, taken on 6 March 1987, also shows the massive Tarantula nebula. It is interesting to compare this photo with figure 91, a shot taken before the supernova made its appearance. PHOTO: JOE CAUCHI

such by professional astronomers throughout the world. When a possible supernova is discovered, large telescopes are turned towards it to get observations. Astronomers are interested in the early stages of supernova explosions, for although we have many records of them fading from maximum brightness there is only one record of observations of a supernova rising to maximum. The supernova search programme is hoping to rectify the situation with early discoveries and notifications. To find out more about the program contact your local society or the Astronomical Society of New South Wales. There are many listings and catalogues of objects which you can observe. We have included over a hundred objects with their positions in the second section of this book, but there are other sources. Perhaps the most famous catalogue of all is that compiled by Charles Messier in the late eighteenth century. Messier was interested in comets, but he was constantly distracted by fuzzy objects which looked like them. Messier would spend many nights observing these to determine if they were moving. To save himself making the same mistake again, Messier made a catalogue of all the objects he found, more than a hundred of them. He numbered them M1, M2 and so on, designations which are still in use. Although Messier did discover a number of comets it is his catalogue that has ensured his fame.

A much more extensive catalogue was published in 1888 by John Dreyer, the 'New General Catalogue of Non-stellar Astronomical Objects' or the NGC. This catalogue listed many thousand galaxies, clusters and nebulae over the whole sky. The catalogue has been revised since, but the NGC number of an object is the official way of indicating it. For instance, the 'Great nebula in

Orion' is also known as M42 as it was the forty-second object in Messier's Catalogue and as NGC 1976 from its position in that list. Other more specialised catalogues have been made, but these are of use mainly to professional astronomers or amateurs with access to large telescopes.

These catalogues give the position of the object in right ascension and declination; they don't give helpful finder charts. To find the object you will need to use a technique similar to that given in chapter 9 for finding stars. Mark the position of the object on your star chart, use the finder to point the telescope in the right direction and put a low power eyepiece into the drawtube. The use of a low power eyepiece is most important when looking for extended objects such as nebulae and galaxies because it will give the brightest image. Keep the low power eyepiece in position and search for the object.

In chapter 2 I mentioned the use of averted vision—that's looking to one side of an object while concentrating your attention on it. This technique is especially useful for deep-sky observing. Another point to note is that a high power will make an extended object like a nebula or galaxy fainter. You will have to balance up the need to see an object clearly with the wish to see more detail. It is not worth increasing magnification to see more detail if that detail then becomes too faint to see.

It is important to remember when you are observing these objects that they will look nothing like the large coloured photographs you may see in books. These photographs are taken with large telescopes; some of them take many hours of exposure to bring out all the details. To the naked eye most galaxies and nebulae appear as faint blobs and smudges, but they are fascinating none the less. As with most things, the more practice you have at deep-sky observing, the better you will become at it and the more detail you will discern. Start observing with some of the easy objects in this book; once you become proficient at finding and viewing these objects, you can move on to the more difficult ones in other publications.

11

Taking astronomical photographs

Observing with the naked eye and taking drawings and notes are very satisfying ways of doing astronomy, but often we would like more permanent or exact records of what we see. Photography fills that need—indeed it was the application of photography to professional astronomy that was one of the biggest steps forward in modern astronomy. Good though it is, there are many things that photography won't do. For example, under good to poor conditions visual astronomy will give much better results than photography as your eye can take advantage of the brief still periods in the atmosphere.

The biggest advantage that photography has over the naked eye is its ability to store light. Our eyes are designed to follow moving objects under good light levels, so they continuously update the information they are receiving. Staring at an object does nothing to increase your view of it. Photographic films on the other hand keep storing the small particles of light, called photons, for as long as they are exposed to them.

To take advantage of astronomical photography you will need very little in the way of equipment, in fact you don't even need a telescope. What you will need is a camera. You probably already have a camera around the house and it may well be suitable. You will need a single lens reflex camera, an SLR. These are made by a number of manufacturers and are readily available. If you need to purchase one, you do not need to spend huge amounts of money for state of the art equipment. In astronomy a camera is mainly used to hold the film in position and to operate its shutter. Electronic metering and program settings may make normal photography much easier, but they are unnecessary for astronomical shots.

One feature your camera will need to have is a B setting on its shutter speed. The B setting, which stands for bulb, lets you keep the shutter open for as long as you hold the button down. You will also need a shutter cable with a locking mechanism to hold the shutter open: you can't do it with your finger as it will shake the camera. You can get a suitable cable for a few dollars at most camera stores.

Other attachments you need will depend upon the type of photographs you wish to take. The best type of photos to begin with are unguided snap shots and piggy-back photos. For these you will also need a normal camera lens. Luckily, the type of lens that comes with most SLR cameras is quite suitable for astrophotography. Camera lenses are known by their focal length; in normal practice these range from about 8mm to 1000mm, with standard lenses being around 50mm. It is a 50mm lens that you should start with.

The focal length of a lens has two effects in astrophotography—the first is the image size. The longer the focal length of a lens, the larger it makes objects appear, and consequently the smaller the field of view seen. It's just like using a high power eyepiece in your telescope: it makes the image appear dimmer, and it exaggerates any small movement of the camera or the object it is photographing. Remember, all objects move across the sky and this motion will show up in your camera. Short focal lengths give small,

83 Some of the large range of colour slide and black and white films suitable for astrophotography. PHOTO: DAVID REIDY

bright images, large regions of sky in the photo and relative immunity from the motion of the heavens. A 50mm lens is a good compromise between all these factors.

The last factor to think about before taking a photo is the film you are going to use. There are many sophisticated films available nowadays and many of them are suitable for astronomy, so the choice is not straightforward. When looking for a film, choose one that gives slides rather than prints. The reason for this is simple. Apart from being cheaper to develop, when a slide film is processed all the laboratory does is put it through the chemicals and then mount the slides. When using a print film it is first processed then printed. In printing, either the operator or a computer decides what your photo will look like. The images of stars on a dark background is usually interpreted as a mistake, and the printed results are rarely very good. It is better to take slides, then you are the only one determining the outcome of the shot; besides you can always get prints made from the slides at a later date. Only buy short films (it is rare to take more than ten shots in any night). A 12 or 24 exposure film will allow you to take it out of the camera sooner, get it processed and see the results.

There is one problem with taking slide pictures. When the laboratory mounts the slides they cut the film into individual pieces. The problem is that they often choose the wrong part of an astronomical film to cut. This usually results in your prize photos being cut in half. One way of stopping this is to ask for the film to be returned unmounted, then buy some slide mounts and cut the film yourself. This takes a fair bit of time to do, and there is an easier way. The first photograph you take on a new film should be of a normal object, your car or your pet. This gives the laboratory a starting point; every so often take another 'normal' slide to give them a reference point. When you take the film for developing, tell the lab that there are astronomical photographs on the roll and that they are to mount every slide. Since Halley's comet most labs have had some experience handling these unusual requests.

As an alternative to colour slide photography many amateurs take black and white photos, then print and process them at home. It really is necessary to do your own developing and printing as commercial firms now charge more for B&W than they do for colour. Processing at home will give you much more control over the final outcome of your photographs, allowing you to do all sorts of tricks with them. The drawback to home processing is the cost of the equipment required to get started and the need for a darkroom. If you already have experience at home photography you might like to apply it to astronomy as well.

When choosing film an important feature to consider is its speed, that is, the rate at which it records light. The higher the film speed the quicker it will build up a picture. The speed of a film is known as its ISO rating, after the International Standards Organization; this was previously known as the ASA rating. Some of the films that are used by amateur astronomers are the Ektachrome range made by Kodak. These come in a variety of speeds with the most popular being the 200 and 400 ISO types. Fuji films and others are also being used; some of these are now available in speeds up to 1600 ISO. But the faster films do have some problems—the higher the speed of a film, the less detail it can record.

84 This twenty minute exposure of the region around the south celestial pole shows the trails stars leave on film if you do not follow their motion. The two faint fuzzy patches on the picture are the Magellanic Clouds. PHOTO: TERRY CLARKE

Another problem with photographic films is known as reciprocity failure. Under normal circumstances the longer you expose a film to the light the more image it makes. If you expose a film for two minutes, you would expect to get twice as much light as if you exposed it for one minute. Unfortunately this doesn't happen—the longer you expose a film the slower the image builds up. On a colour film this leads to colour imbalances. Because the various layers of a film have slightly different reactions to reciprocity failure different colours come out. On Kodak films this is usually a green background to the sky; on other films it shows as other colours.

There are a number of methods used to overcome reciprocity failure and to increase the speed of a film. The one that is most accessible to amateur astronomers is gas hypersensitisation. Film treated in this method, usually called hypered film by amateurs, has been kept in a mixture of hydrogen and nitrogen gases at about 50°C for a few days. This treatment will reduce reciprocity failure to a reasonable level and about double the film speed. Unfortunately hypered film will deteriorate quickly if not stored properly; this usually involves keeping the film in the refrigerator in its original container. Another problem is that hypered film must be processed very quickly after it has been exposed. That means that it shouldn't remain in the camera for more than one night. When you have finished with the film you will need to send it to be processed promptly.

Since Halley's comet, a number of companies have sprung up which specialise in supplying hypered film and other services to the astrophotographer. These companies operate by mail order, supplying the film and arranging for it to be processed afterwards. The cost of hypered film is not a lot more than normal film, though it is usually only available in small lengths. Your local astronomical society will probably have contact with one or two of these companies and can put you in touch with them.

To begin your explorations in astrophotography start with some simple snapshots. To begin, you will need a tripod to hold your camera steady, though you may be able to use your telescope mount for this. Load your camera with some 400 ISO slide film, don't forget that first photo of the cat, and take your camera out to a dark site. Selecting a good site for photographing is more important than for visual observing; any extra light in the area will quickly expose the film in your camera, rendering any picture a dull foggy mess. Choose a site that is in deep shadow if you can't get completely away from lighting—there is nothing wrong with putting the camera close to the ground to achieve this.

Point your camera at any place in the sky, though the region around the south celestial pole is the most interesting, and make sure that there are no street lights or other lighting in the frame. Set the shutter speed to B and using the shutter release cable keep the shutter open for one minute. Easy isn't it? Now do the same thing with the shutter open for fifteen seconds, thirty seconds, two minutes, five minutes, fifteen minutes, half an hour, even longer if you can stand sitting in the dark for that long. Remember that the film in the camera will record all light, so you can't use your red torch once you've got the shutter open.

When taking photographs it is very important to take copious notes, otherwise how will you repeat your successes and avoid remaking mistakes? In addition to the usual record of universal time and the object being photographed you should also record the lens you are using on the camera, the film you've got in and the length of the exposure. It is also an idea to give each photo a unique number that you can also record on the slide mount or print. One system is to use a number such as 200187-04 for each photo; the first numbers are the date of the photo and the last digits the number of the photo for that night. This example would be the 4th photo on 20 January 1987. In this way you should be quickly able to find the record for any of your works.

Once you have had the film processed you will be able to see a number of the fundamental features of astrophotography. To view your slides you do not need a slide projector—there are many hand-held slide viewers available, and these are much cheaper, and easier to use than having to set up a projector every time you want to see your masterpieces. Have a look at your fifteen second shot. You should be able to see some faint star images against a dark sky. The thirty second shot will show more stars, the one minute shot even more detail. The two minute shot will not show many more stars than the one minute one, but you will see another effect. Instead of the point-like images on the shorter exposures, the two minute shot shows the stars as short dashes. These dashes are caused by the rotation of the Earth and the consequent motion of the stars in the sky. In the longer exposures these dashes will turn into arcs of circles. If you have the SCP in the shot then you will see that all the star trails have it as their centre.

Another effect that will show up in the longer exposures is fogging. The sky anywhere near a city has a certain amount of light in it, although this may be too small to be seen with the naked eye. Your film will pick up this light and record it; eventually there is enough to make a background colour to the photograph. This sky brightness will limit the length of exposure you can make with later techniques. The only way to

Table 11.1 Optimum exposure times for astronomical photography

Focal length in millimetres	Exposure time in seconds
8	250
18	110
25	80
35	60
50	40
70	30
100	20
150	13
200	10
300	7
400	5

Note: This table will allow you to estimate the length of time you can photograph before the motion of the sky will interfere with the image. The times are an average for the sky—if you are shooting close to the celestial equator, you must shorten the time, if your photographs are of the polar regions, you can lengthen them.

If you want to calculate the time for yourself, use the following formula:

$\text{Time} = 1000/(f \times \cos\delta)$

Where Time is in seconds; f is the focal length of the lens in mm; δ is the declination of the object you are photographing.

cure it is to go to a country location and photograph from there.

As you can see from your photographs, the motion of the sky is likely to be a problem for exposures of any length of time. Table 11.1 gives you some idea of how long you can expose your film without the movement becoming noticeable. The times given are for the average star; you will have to allow for shorter periods if photographing close to the celestial equator, but you can allow longer if shooting near the pole.

Taking short exposures is one way of solving the problem of the sky's movement, but there is a better way. If you have an equatorial mount you will know that you can follow an object across the sky by moving it in just one direction. This facility can also be used to keep your camera pointing the right way. The simplest way of using your mount is to do 'piggy-back' photography. In this technique your camera uses its normal lens for the photograph, but you mount it on the back of the telescope.

To attach your camera you will need a mounting device. Such a device can be purchased quite cheaply from most telescope shops. When choosing, make sure that it fits the telescope properly, and that it is firm enough to hold the camera without wobbling. The camera will attach to such a device through the threaded hole in its base, the one used for fitting the camera to a tripod.

When you attach the holder to the telescope put it near the front end. This will make sure that the telescope itself doesn't appear in your photos. If you have a Newtonian, make sure that you can still get your eye to the eyepiece without interfering with the camera. Because an equatorial mount is delicately balanced you will find that adding a camera will upset this. The solution is to rebalance the telescope the same way as you did when you first purchased it. Of course, if you remove the camera, the telescope will again be unbalanced, so it is often easier to leave the camera on the telescope while you are observing, whether you are taking photographs or not.

If you intend to do photography with your telescope you will need to be extra careful when you set up the mount. Make sure that it is pointing directly south and that the angle for your latitude is set correctly. Doing this carefully when you are preparing will make the job a lot easier when you are photographing. With the camera attached to the telescope and the mount aligned you are almost ready to start. Select the area of sky that you would like to photograph and point the telescope and the camera in that direction. If you have a finder scope with crosshairs in the eyepiece, remove the eyepiece from it and put it in the telescope in place of the regular one. Having crosshairs will make the job a lot easier.

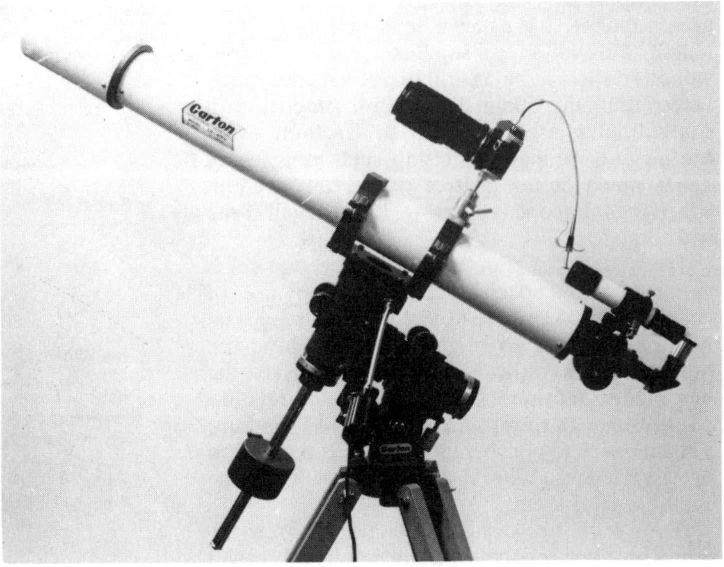

85 This refracting telescope has been equipped with a mount for piggy-back camera use. The camera, its shutter held open by the release cable, moves with the telescope while it is kept centred on the object being photographed.
PHOTO: DAVID REIDY

Now select a bright star in the area and centre the crosshairs, or the field of view of the telescope, on it. Look through the camera to make sure that the region you want is still in view. If you have a drive mechanism for your mount now is the time to turn it on. Once you are all set, turn off any lights around and, using the shutter release cable, start the exposure. Once you are going you will need to constantly monitor the view in the eyepiece. The motor drive, if you have one, will keep the crosshairs fairly close to the bright star you selected, but you may need to make minor adjustments from time to time using the slow motion controls.

If your mount does not have a motor drive then you will have to do a lot more of the work. Using the control that adjusts the telescope's right ascension, you will need to track the star through the sky. This will take a little practice, but after a time you will get a feel for just how much you need to turn the control. Occasionally you may need to make an adjustment of the declination control as well.

The more accurately you do the tracking, or the more accurately you make the necessary corrections, the better the resulting photograph will be. Make it easy on yourself to begin with, take the photographs using the camera's 50mm lens and use areas near the pole where mistakes are harder to make. As you progress, you can move onto longer exposures, lengths up to the limit imposed by the sky brightness. Once you are confident, try using a lens of longer focal length. This will give a smaller field of view, but will often show nebulae and bright galaxies much better than the 50mm lens. With practice you can probably use lenses up to 1000mm using the piggy-back method. Using such long lenses needs much better control of the direction in which the telescope is pointing, but this will come with practice.

There are two objects in the sky on which you need to use special techniques. They are, of course, the Sun and the Moon and they're special because they are so bright. Photographing the Sun is almost as difficult as observing it. Even the new high tech cameras with titanium shutters are not fast enough to capture any detail on the Sun. The only way to photograph the Sun directly is to use a very dense neutral density filter, or maybe even a few of them.

If you have been using projection to view the Sun with your telescope, there is a much easier method of obtaining a photo. Take a photograph of the projection screen. The advantages of this method are quite clear. Not only do you not have to worry about filters and exposure times, but you can take the photograph as you would of any normal subject, using the built-in features of your camera to make this easier. The only thing you need to worry about is focusing the camera and getting it positioned correctly. It is important that you get the camera as square to the projection screen as possible. This is complicated by the fact that the camera itself sometimes gets in the way of the light from the telescope. I find that putting the camera as close as possible to the eyepiece of the telescope usually gives the best results.

Table 11.2 Suggested exposure times for photographing the moon

f/ of lens	Exposure in seconds	Eclipse
1.4	1/8000	1/2
2.0	1/4000	1
2.8	1/2000	2
4	1/1000	4
5.6	1/500	8
8	1/250	15
11	1/125	30
16	1/60	60
22	1/30	2 min
32	1/15	4 min
45	1/8	8 min
64	1/4	16 min

Note: [a] It is best to bracket your exposures, that is, take one photo at the recommended speed, one at one speed faster and a third at one speed slower. This usually ensures that at least one of the shots will be good. The times given for eclipse photographs are very approximate. They are based on an average eclipse; if the eclipse is very dark you will need to increase the length of exposure, if it is a particularly light eclipse you will need to shorten it.
[b] These exposure times are calculated for 400 ISO film and for a one quarter moon. If you are using a different speed film, use the table below to multiply the time given. For a different phase of the moon, multiply the time given by the number below.

Speed of film	Multiply time by
64	8
100	4
200	2
400	1
800	1/2
1600	1/4

Phase of Moon	Multiply time by
Thin crescent	4
Thick crescent	2
Quarter	1
Gibbous	1/2
Full	1/4

The Moon is the other object that can be photographed quite easily. Because of the large amount of light coming from it, exposures of the Moon are quite brief so you don't need to guide for them. I find it is easiest to use a tripod for most exposures, but even this is not necessary. The Moon is a small object, so using a telephoto lens on it will give best results. In table 11.2 I give some suggested exposure times for the Moon.

When you are doing astronomical photography, it is best to use a technique called bracketing to make sure that your photos come out. Bracketing involves taking a number of photographs of an object instead of just one. Firstly you calculate the exposure that you think is necessary for the object and then take the shot. You then double the time and take a second shot, lastly you halve the first time and take a third shot. This should ensure that at least one of the photographs will be what you want. This method does use up a fair bit of film, but often one of the bracketed exposures will be better than your first guess.

Once you have mastered the simple techniques required for piggy-back photography you will be able to try some of the more complicated techniques in astrophotography. The lens or mirror system of your telescope can be used just like a camera lens. In fact, some of the best long telephoto lenses are really maksutov reflecting telescopes. To use this method of taking photos, known as prime focus photography, you will need a special camera adaptor. Suitable adaptors can be bought at most telescope shops and even some photography stores.

The adaptor you need for prime focus photography, and also for eyepiece projection photography as we shall see, is a short tube which attaches to your camera at one end, and fits into the eyepiece holder of the telescope at the other. The part that fits into the telescope is standard, either 24.5mm or 1.25″ in diameter; the other has a standard fitting known as a T-ring. Nearly every camera manufacturer uses a different method of attaching lenses to their products, so you will also need to buy a T-ring adaptor for your camera.

To do prime focus photography with your telescope you will need to do much of the same work as you did for piggy-back photos. You will need to balance the telescope and make sure that the mount is correctly set up and aligned. You must also make sure that the finder scope is pointing correctly. This is more important because you will use the finder scope to guide the telescope while you are taking the photo. In fact, many amateurs who do this type of photography with large telescopes use a second smaller telescope to help them guide. Often these guide scopes are small refractors of about 60mm aperture. When you purchase a larger telescope in the future this is a perfect application for your first instrument.

Taking prime focus shots is simply a matter of pointing the telescope in the correct direction, swapping the eyepiece for the camera and its adaptor and focusing the telescope using the camera. Then you track the object by viewing through the finder or guide scope. Because of the huge focal length of telescopes compared with most camera lenses, accurate tracking is most critical. It is at this time that a telescope drive is most useful as it will do a lot of the hard work for you. If you don't already have one, and you intend to do a lot of photography, now is the time to consider the purchase.

The last of the amateur astrophotography techniques is called eyepiece projection. This uses both the telescope and an eyepiece as the lens system for the camera. Once the telescope is set up for prime focus photography it is a simple

86 This camera has been attached to the telescope for prime focus or eyepiece projection photography. The small adaptor tube which connects the camera to the drawtube of the telescope can either be used empty for prime focus phtography, or with an eyepiece for projection. PHOTO: DAVID REIDY

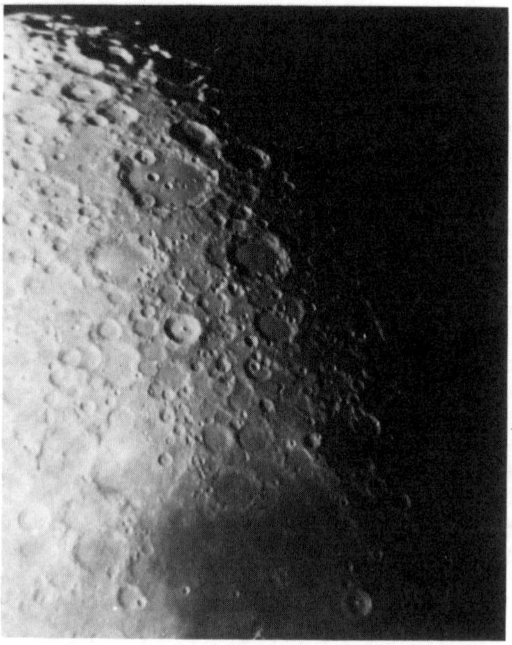

87 This photograph of the Moon shows what can be achieved with eyepiece projection photography. PHOTO: KEN WALLACE

task to put an eyepiece into the camera adaptor. Fit this to the telescope and focus the telescope while looking through the camera. Expose in the usual method using the shutter release cable. The combination of the eyepiece and the telescope gives a very large magnification for the photograph, and a large number of problems for the photographer. Tracking is critical and the low light level that results from the combination means that long exposures are necessary for most shots. If you want to try this method, I suggest that you get some hypersensitised film and try only bright objects.

It is difficult to calculate the exposures for eyepiece projection as they depend upon many factors, the focal length of the telescope, the focal length of the eyepiece and the distance from the eyepiece to the film, just to name a few. The best teacher is experience, so you will need to experiment. Choose an easy object like the Moon and do some exposures ranging in time from 1 minute to half an hour.

Eyepiece projection will really test your skills as a photographer, and is definitely not recommended for the beginner. Try the simpler techniques first and join the local astronomical society. There are always other members who will give you a hand getting started.

12

The modern objects

The frontiers of professional astronomy are always expanding, new objects are being discovered, new explanations for phenomena are being devised, old ones are being discarded. Although you can't see most of the objects described in this chapter even with a large amateur telescope, no book on astronomy would be complete without a brief discussion of them.

There is one point to be remembered though, all of this astronomy is on the forefront of science and so the current explanations may well be wrong. There was a time when the scientific establishment said that the Earth was flat, that the Sun and planets all orbited the Earth. Of course, we now know that these ideas were quite mistaken. Our explanations of new phenomena may also be quite wrong and in a few years new techniques will give us fresh solutions to these puzzles.

Much of recent astronomy has been done not with the light by which we see, but with other forms of electromagnetic radiation. The light by which you are reading has a wavelength of about 500nm, about one two thousandth of a millimetre, between each crest. There are other waves about us with longer and shorter wavelengths. The waves that carry television and radio have wavelengths of a few metres. The x-ray and γ-ray waves of radiation from nuclear bombs are only a few millionths of a millimetre long. Astronomers now have methods of observing both in radio waves and also x-rays and very recently γ-rays.

It was after the second world war, when an engineer called Carl Janskey was working on radio reception, that astronomers first started observing using things other than light. Janskey noticed that a source of interference in his work seemed to move across the sky; he also noticed that it moved with the sidereal period of 23h56m04s rather than 24 hours. Very quickly this object was identified as Jupiter and the study of radio astronomy began. Nowadays as much astronomy is done using radio telescopes as using optical ones. Australia was a leader in the early days of radio astronomy and even today a great deal of very valuable research is carried out by universities and the C.S.I.R.O.

With the coming of space flight, astronomers were able to get their instruments above the protective layer of atmosphere which surrounds tht Earth. In space all sorts of radiation are present, much of which never reaches the Earth's surface because our air absorbs it, keeping life here safe. In space the very high energy x-rays and γ-rays can be studied, giving scientists an insight into some strange processes. In the last few years instruments such as the Einstein x-ray Satellite have shown us objects that we never knew existed, objects that astronomers are still at a loss to fully explain. These are the modern equivalents of the new objects Galileo saw with the first optical telescope. A whole new field of science.

Perhaps the first of the modern astronomical objects was observed in 1967. At this time Jocelyn Bell and Martin Hewish were using a radio telescope at the Cambridge University Observatory. On a trace from the telescope Bell discovered some pulses that showed remarkable

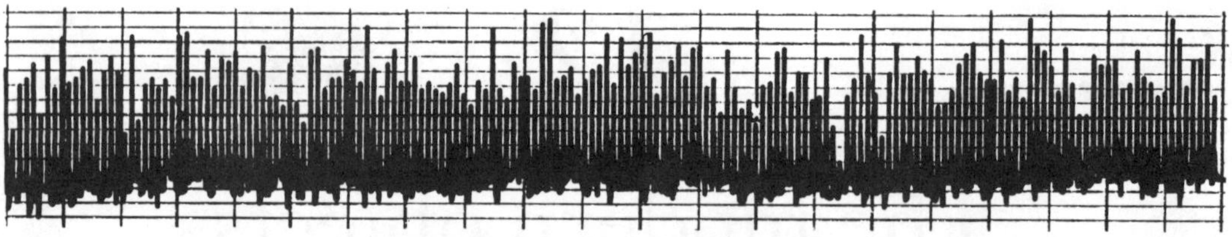

88 This trace of the signals from pulsar PSR 1641-45 was made on 29 November 1981 using the Parkes radio telescope operated by the CSIRO. The pulses from this pulsar are about 0.5 seconds apart.

regularity. At first these were dismissed as being either instrumental errors, or perhaps ignition noise from a passing motor car. A few nights later these pulses were again observed in the same position of the sky; it was unlikely that these were of terrestrial origin.

A few more observations gave a position of 19h19m36s +21°47′. The pulses were also shown to have a very regular period, 1.337301s, much more precise than any other astronomical object, and of similar precision to the best atomic clocks then available. At once theories arose that this object was a kind of beacon for extra-terrestrial beings to use for navigation. The truth was quickly deduced though: these pulses were caused by the remnant of a star, the star was called a pulsar. Within a few years more pulsars had been found, and many more are found each year.

We saw in chapter 9 that stars have a life cycle; they are born, live, and eventually die. Some stars collapse so much that the atoms that form them fall apart. Instead of remaining as atoms of hydrogen, carbon or such, the atoms turn into collections of neutrons, a sub-atomic particle. Eventually, the whole star is made of these particles; it becomes a neutron star.

These stars have some very strange properties. Because it has collapsed so much, a neutron star is very dense: a cupful, if you could get such an amount, would weigh many thousands of tonnes. The neutron star itself is only a few kilometres in diameter, its surface consisting of a layer of iron with a neutron core beneath it. When the original star collapsed to form the neutron remnant, its magnetic field remained. A neutron star has the magnetic field of a normal star squeezed into a few kilometres so the field becomes incredibly strong. The rotation of the original star is also preserved in the contraction, though because the star is much smaller, it is quicker—a new pulsar may rotate as many as five hundred times a second.

It is the quick rotation and strong magnetic field of a pulsar which cause the mysterious pulses. When electrons, another type of sub-atomic particle, enter the magnetic field of a pulsar they give off energy, in light waves, radio waves and other forms of radiation. This emission of energy is strongest at the magnetic poles of the pulsar. From the polar regions energy is radiated out into space like the beam of a searchlight. The rotation of the pulsar sweeps these beams through space like a lighthouse sweeps its beam of light. From our vantage point on the Earth, this appears to us as the pulsar flashing on and off each time the beam sweeps past us.

The best researched of all the pulsars we now know of, is the Crab pulsar. The object is found near the centre of the Crab supernova remnant in Taurus. The nebula is the remains of the star's outer layers. The pulsar, which now appears as a 16th magnitude star, is the remains of the star's core. The Crab pulsar has been recorded flashing in visible light, infrared, ultraviolet and even x-rays as well as radio, giving us great insight to the working of these objects. Since the Crab pulsar identification a number of other pulsars have been related with nebulae from supernovae, but many more haven't.

Another phenomenon that was discovered with a radio telescope was the bright star which appears to be a radio source. In 1960 a survey of radio sources was being made by Cambridge University, when a radio source was identified with an object that looked like a bright star. As this was the 273rd object in the catalogue, it was called 3C 273, 3C for the third Cambridge Catalogue. This object and two others, 3C 48 and 3C 147, were thought to be radio stars and much research was undertaken to study them. Very

89 The Crab pulsar, perhaps the most studied of all radio astronomy objects, contains a pulsar at its centre which causes the nebula to shine. Observations of this pulsar go back to 4 July 1054 when Chinese astronomers saw a star explode as a supernova. The nebula and pulsar are the remnants of this explosion.
PHOTO: JOE CAUCHI

quickly these objects became known as QSOs, quasi-stellar radio objects; this has now become 'quasar'. One of the first tasks done was to get spectra for the objects; when this was done a strange result was seen. The spectral lines matched no known element!

In 1962 Maarten Schmidt was studying 3C 273's spectrum when he noticed that the lines could be caused by hydrogen, if they were shifted by 16% towards the red. The movement of spectral lines is caused by the movement of an object away or towards the observer, as we saw in chapter 10. Until this time the largest red shift observed had been only a few percent from very distant galaxies. This 16% shift meant that 3C 273 was moving away from the Earth at 48 000 kilometres per second. If Hubble's Law was correct then this object was close to 3 thousand million light-years distant. Having an object at such a great distance created problems; if the quasar appears in the sky with a magnitude of 12.8, then the object itself must be incredibly bright. Much brighter than anything yet seen, with an absolute magnitude of −27: that's over four billion times brighter than the Sun.

Because the quasar was so bright, astronomers were at a loss to explain the cause of such energy. Explanations ranged from the denial ('the measurements must be wrong'), to the ridiculous ('these are wormholes connecting our universe to another'). After many years of diligent research, and the discovery of many more quasars, a seemingly correct explanation has emerged.

As we know, the further we look into the distance, the further back in time we look. For 3C 273 this is 3000 million years, back to the time that life was just starting on the Earth. Some of the more recently discovered quasars have distances corresponding to 10 000 million years, before the Sun had even formed. Conditions in the universe were probably very different to those that we now see, and this helps to explain what a quasar is. Ten thousand million years ago the galaxies we see today were just beginning to form; this is what quasars probably are, the dense bright cores of new galaxies. It is important to stress that this is a new theory and may be found to be quite wrong in the future, but for the moment it is the best we have. Astronomers throughout the world will be doing more research to find the truth.

The most exciting object of modern astronomy is the 'black hole'. Black holes are used nowadays to explain many of the high energy phenomena observed in modern astronomy. It is important to realise that, although black holes are used to explain almost anything, we have still not observed one. I can't help feeling the use of black holes is a little akin to the phrase 'Here there be monsters' that appeared on old maps. We don't know exactly what is going on, or what a black hole does, so we can use them to explain anything we want.

Although their use in astronomy is quite recent, the idea of a black hole is not new—Isaac Newton mentioned the possibiltiy of their existence and Einstein formalised their properties in 1915. Since that time much more work has been done, mostly by theoretical physicists and mathematicians, to exactly define the properties and effects of a black hole.

As you know, every object with mass causes gravity, and this gravity affects all other objects, even light rays. The denser and more massive an object is, the greater its gravitational field. If you get the right combination of density and mass, then you can create a gravitational force so strong that nothing can escape from it, not even light. This is a black hole, an object that everything falls into, but nothing escapes from.

90 This photograph shows the United States' space station Skylab a few hours before it crashed back to Earth in July 1979. PHOTO: KEN WALLACE

It's like a hole in the universe. As an object falls into a black hole, it gives off huge amounts of energy, energy which appears as radiation, both visible light and all the other forms. It is this radiation that should be detectable from Earth.

Black holes are thought to form when a star collapses too far. We have seen that most stars end their lives as white dwarves. Some stars collapse too far to be dwarves and form neutron stars. Very massive stars are so heavy that they can't stop even at the neutron star stage; they continue to collapse because there is no force strong enough to stop them. It is these objects that form black holes. As they pull matter into them from their surroundings, they become more and more massive, and pull more and more material into them. Some astronomers say that at the centre of some galaxies there are black holes that eat whole stars at a time, and will gradually devour their whole galaxy. It is perhaps fortunate that there are no black holes anywhere close to us.

At present there are a number of candidate objects for black holes, but our instruments are still too poor for a definite identification to be made. In coming years as more and more sophisticated devices come into use, and telescopes are put into space to get away from the effects of atmosphere, scientists may get enough data to positively identify a black hole.

13

About the charts and objects

The objects in the list following this chapter have been selected from the many available to the southern observer. Since the majority of astronomy books published have been written for the northern hemisphere, the southern skies have been sadly neglected. With a few notable exceptions, the objects in this book have declinations south of +20 degrees, therefore all may be observed comfortably from the southern hemisphere.

The objects have been listed in order of right ascension to enable you to plan a night's viewing. It is not intended that you start reading from the first object to the last, but rather use the guide as a companion to your observing.

When planning an evening's observing, select objects that will be on, or within a few hours either side of the meridian. Table 9.2 will assist you here. The idea is to view objects when they culminate, that is when they are highest in the sky; this decreases the amount of atmosphere through which you are looking. Light pollution, the curse of the astronomer, will be minimised in urban areas if objects are viewed at their greatest altitude.

If you are a city dweller and are planning a Saturday night's excursion to the country, concentrate on the fainter nebulae and galaxies that would normally be beyond your grasp from urban skies. Exclude objects such as double stars as they are observable at any time—light polluted skies do little to detract from their beauty. In fact, the Moon, planets, double stars and the brighter nebulae and clusters all may be observed from the city, so take heart, you do not have to live on a farm to enjoy astronomy.

Always spend some time before observing to list the objects you intend to view. Your selection should contain some new and challenging objects along with the familiar ones. This will prevent you from habitually viewing the same old bright objects over and over again, eventually making the hobby boring.

Experiment and use dark skies to push your telescope to its limits. That tiny, insignificant magnitude 12 smudge in your eyepiece may not look particularly impressive. If, however, you consider it as an 'island universe' receding from you at 1000 kilometres per second, composed of more than a hundred million suns, its light having taken 50 million years to reach you, the whole universe takes on new meaning. With awe you are humbled into quiet solitude—the immensity and grandeur of it all, away from familiar city noises, is indescribable. Spend a little time away from the eyepiece and take in the heavens with the naked eye, listen to the quiet; it is these interludes that provide one with the opportunity to ponder the universe.

The objects listed have been selected as a sampling of the finest the southern skies have to offer. Much diversity within each class of object has been sought. All are visible in small instruments, and hopefully they will whet your appetite, beginning many enjoyable years of amateur astronomy.

Do not underestimate the capabilities of your instrument, no matter how small. For my eighth birthday I received a humble 40mm refractor; I used and abused this instrument for the next eight years until, at the age of sixteen, I was able to save for a 150mm reflector. I consider those

91 This photograph of the Large Magellanic Cloud and Tarantula nebula shows some of the interesting objects in the region near the south celestial pole.
PHOTO: KEN WALLACE

early years as an apprenticeship when I learned much about the heavens. My fondest memories come from that era, when I saw for the first time many of the objects listed in this book.

Many amateurs have made significant discoveries with the unaided eye, binoculars and small telescopes. Both Galileo and Messier worked with instruments that were smaller and inferior in quality to those the average beginner has today. Of course, the larger the telescope, the more you will see, and the brighter the objects will appear, but do not be disillusioned. Inexperience and great expectations have seen many a new telescope shelved before its owner learned the arts of using the instrument. Allow your eyes to become dark adapted, use averted vision on faint objects and give yourself the time to master the art of seeing. Telescopes are not 'user friendly': they require patience and practice to obtain the most from them. Only upgrade to a larger instrument when you consider that you have completely exhausted the possibilities of your current one. Most of the observing I do from my suburban backyard is with an 80mm refractor; it is easily set up in a few minutes, and is adequate for most objects in light polluted skies. A 250mm reflector is reserved for my trips to the mountains and rural areas where I observe and photograph deep-sky objects.

Rather than include photographs in this book taken by professional observatories, we have selected shots taken by amateurs to show the type of work you too can do. Admittedly these amateurs have been involved in astrophotography for many years, but the equipment they use is probably no more sophisticated than your own. Provided that you have an equatorial mount that is set up correctly, a motor drive, a camera and patience, there is no reason why you cannot duplicate some of these shots. In fact, I have seen excellent photographs of the Milky Way taken with a camera mounted piggy-back on a telescope without a drive, the photographer patiently guiding the camera with the telescope's slow motion controls.

To describe where an object is, its size or separation, I have used angular measurements. At first these may seem a little confusing, but with practice and familiarity with your instrument, this will quickly change. To understand distances expressed in this fashion all you need to remember is that there are three hundred and sixty degrees (360°) in a circle, sixty minutes (60′) in a degree and sixty seconds (60″) in a minute. You will also need a yardstick to gauge these distances. Look at the Southern Cross with your finder scope and make a note of the area it covers. The distance from α Crucis to γ Crucis is 6°, so when an object is described as 5° from another you will have some idea of how far that is in your finder.

You will also have to calibrate your telescope.

92 The Trifid, object 78 (lower) and Lagoon, object 81 (upper) nebulae are two fine examples of objects that can be seen clearly in amateur telescopes. PHOTO: KEN WALLACE

The Moon is half a degree in diameter, or 30 minutes of arc. Use the various eyepieces you have to observe the Moon and note how much of it fills the field; make a note of this. For example, if the Moon just fills the field of your 25mm eyepiece, you will know that whenever you use this eyepiece it will give you a 30′ field of view. Double stars, because they are so close together, are measured in seconds of arc. To get a feel for these, look at the double star μ Crucis, object 59; its elements are separated by 34.9″ of arc. Chapter 9 describes an accurate method of measuring the exact separation of binary stars.

All the objects given are presented in similar formats. Firstly the object's catalogue number is given, for non-stellar objects this is its NGC number. This is followed by any other catalogue numbers, Messier's objects for example, and lastly its popular name and the type of object. Double stars are given the name of their discoverer. Most charts show these doubles with a symbol to represent the discoverer, but for the sake of simplicity I have used the discoverer's name in the descriptions. For example, charts give James Dunlop's stars a capital delta, Δ, as an identifier, and John Herschel's an 'h'.

If more than one object is located within a field, or is nearby, I have grouped them together under one heading. The constellation that the object or objects are in is then given along with a position in right ascension and declination. The RA and Dec. are always given for Epoch 1950.0 and the charts are all drawn for this epoch. If your telescope is equipped with setting circles, and set correctly, you will be able to dial up any object. Learning to use your setting circles properly is a distinct advantage, especially in light polluted skies where locating an object with your finder may be difficult. You will find more on using setting circles in chapter 6. Where more than one object is listed, only the RA and Dec. of the first are given. To find the exact position of the other objects refer to the lists in table 13.1.

The charts themselves have been drawn to help you find each object. Where possible much of the neighbouring region of sky has been shown, so that you can star hop from the brighter stars. The objects are identified on each chart using the symbols given at the beginning of the charts. As an additional aid to finding the constellations, aspect maps for Spring, Summer, Autumn and Winter have been included. All the charts have been drawn with north at the top so that after locating the chosen constellation you can simply orientate the book and identify the stars you see.

There are just over a hundred objects in this book, and though they will keep you busy for a fair while, they will eventually run out. At this time you may care to acquire a good quality star atlas. These come in many forms giving the position of many more objects, but as they don't give descriptions you should be prepared to start your own investigations. Also, there are catalogues which give information and positions of objects and which are available separately from the atlases. It can be expensive to start an astronomical library and advice should be sought from other amateurs as to what is good value. By joining an astronomical society, you can often borrow books from the library before buying them. A society also offers the chance to meet other amateurs, to compare notes and talk with people who share your interests.

Table 13.1: Objects by class

RA	DEC.	CATALOGUE NUMBER	OBJECT No	CHART No
Double and multiple stars				
02h56m24s	−40°30′	Theta Eridani	5	4
04h33m00s	−62°56′	Herschel 3670	8	2
04h49m48s	−51°48′	Iota Pictoris, Dunlop 18	9	4
05h12m06s	−08°15′	Beta Orionis, Rigel	10	5
05h32m36s	−06°02′	Struve 747	21	5, 5a
05h32m48s	−05°25′	Theta Orionis	15	5, 5a
05h32m54s	−04°52′	42 Orionis	17	5, 5a
05h33m00s	−05°56′	Iota Orionis	20	5, 5a
06h26m24s	−07°00′	Beta Monocerotis	25	7
06h28m36s	−50°12′	Dunlop 30, Rumker 65	26	6
06h56m42s	−28°54′	Epsilon Canis Majoris	28	7
07h25m30s	−11°27′	Struve 1097	30	7
07h33m48s	−14°23′	Struve 1120	34	7
07h34m18s	−14°22′	Struve 1121	35	7
07h36m48s	−26°41′	k Puppis	36	7
08h07m54s	−47°12′	Gamma Velorium, Dunlop 65	39	6
10h17m12s	+20°06′	Gamma Leonis	42	12
10h30m48s	−46°45′	Herschel 4330, t Velorium	44	11
10h31m24s	−46°43′	Herschel 4332	45	11
12h23m48s	−62°49′	Alpha Crucis, Acrux	56	10
12h51m36s	−56°54′	Mu Crucis, Dunlop 126	59	10
14h36m12s	−60°38′	Alpha Centauri	66	10
14h38m24s	−64°46′	Alpha Circini, Dunlop 166	67	10
15h16m42s	+01°57′	5 Serpentis, Struve 1930	69	16
16h26m18s	−26°19′	Alpha Scorpii, Antares	71	17
17h59m18s	−23°02′	HN40	80	17
18h02m54s	+02°31′	70 Ophiuchi	83	19
18h53m48s	+04°08′	Theta Serpentis, Struve 2417	90	21
19h19m00s	−44°33′	Beta Sagittarii, Dunlop 226	92	23
19h35m00s	−16°21′	Epsilon Sagittae	93	21
19h37m12s	+16°27′	HN84, Hh630	94	21
19h48m42s	−55°06′	Dunlop 227	95	23
21h16m18s	−53°40′	Theta Indi, Herschel 5258	98	23
22h28m42s	−32°36′	Beta Piscis Austrini, Dunlop 240	102	1
22h37m00s	−28°36′	Herschel 5356	103	1
Open star clusters				
03h43m54s	+23°58′	M45, the Pleiades	6	5, 5c
05h33m00s	−04°24′	NGC1981	19	5, 5a
06h04m18s	+24°06′	NGC2158	24	5
06h05m42s	+24°20′	NGC2168, M35	23	5
06h44m54s	−20°42′	NGC2287, M41	27	7
07h16m36s	−24°52′	NGC2362	29	7
07h34m18s	−14°22′	NGC2422, M47	32	7
07h39m18s	−14°42′	NGC2437, M46	31	7
07h50m30s	−38°25′	NGC2477	37	6
07h59m42s	−60°44′	NGC2516	38	8
08h37m30s	+19°52′	NGC2632, M44	40	9
10h31m30s	−57°58′	NGC3293	46	10
10h40m24s	−64°50′	Mel 101	48	10
10h41m00s	−64°08′	I.2602	47	10
11h03m24s	−58°24′	NGC3532, Dunlop 323	50	10
11h34m00s	−61°19′	NGC3766, Dunlop 289	54	10
12h50m36s	−60°05′	NGC4755, Dunlop 301	58	10
17h36m48s	−32°11′	NGC6405, M6	72	17

Table 13.1 continued

RA	DEC.	CATALOGUE NUMBER	OBJECT No	CHART No
17h50m42s	−34°48′	NGC6475, M7	76	17
18h01m48s	−22°30′	NGC6531, M21	79	17
18h01m48s	−24°20′	NGC6530	82	17
18h16m00s	−13°48′	NGC6611, M16	84	20
18h25m06s	+06°32′	NGC6633	87	21
18h48m24s	−06°20′	NGC6705, M11	89	20
Globular clusters				
00h21m54s	−72°21′	NGC104, 47 Tucanae	2	2
13h10m30s	+18°26′	NGC5024, M53	61	15
13h23m48s	−47°13′	NGC5139, Omega Centauri	63	14
13h43m00s	−51°07′	NGC5286, Dunlop 388	65	14
15h16m00s	+02°16′	NGC5904, M5	68	16
16h20m36s	−26°24′	NGC6121, M4	70	17
17h36m48s	−53°39′	NGC6397, Dunlop 366	74	18
17h46m48s	−37°02′	NGC6441, Dunlop 557	75	17
18h33m18s	−23°57′	NGC6656, M22	88	20
19h06m24s	−60°04′	NGC6752, Dunlop 295	91	22
19h51m30s	+18°39′	NGC6838, M71	96	21
21h30m54s	−01°04′	NGC7089, M2	99	24
21h37m30s	−23°25′	NGC7099, M30	100	24
Planetary nebulae				
04h12m06s	−12°52′	NGC1535	7	5
05h31m30s	+21°59′	NGC1952, M1	13	5
07h39m18s	−14°36′	NGC2438	33	7
10h04m54s	−40°11′	NGC3132	41	11
10h22m24s	−18°23′	NGC3242	43	13
11h47m48s	−56°54′	NGC3918	55	10
21h01m24s	−11°34′	NGC7009	97	24
22h27m00s	−21°06′	NGC7293	101	1
Diffuse nebulae				
05h32m54s	−05°25′	NGC1976, M42	14	5, 5a
05h33m00s	−04°52′	NGC1977	18	5, 5a
05h33m06s	−05°18′	NGC1982, M43	16	5, 5a
05h39m54s	−69°04′	NGC2070, Dunlop 142	22	2
10h43m06s	−59°25′	NGC3372, Dunlop 309	49	10
17h58m54s	−22°30′	NGC6514, M20	78	17
18h01m36s	−24°20′	NGC6523, M8	81	17
18h16m00s	−13°48′	IC4703	85	20
18h18m00s	−16°12′	NGC6618, M17	86	20
Galaxies				
00h12m30s	−39°50′	NGC55, Dunlop 507	1	1
00h45m06s	−25°34′	NGC253	3	1
00h50m00s	−73°30′	Small Magellanic Cloud	12	2
05h20m00s	−73°30′	Large Magellanic Cloud	11	2
11h16m18s	+13°13′	NGC2623, M65	51	12
11h17m42s	+13°53′	NGC3628	53	12
11h17m48s	+13°17′	NGC3627, M66	52	12
13h02m24s	−49°13′	NGC4945, Dunlop 411	60	14
13h22m24s	−42°45′	NGC5128, Dunlop 482	62	14
13h34m18s	−29°37′	NGC5236, M83	64	14
Miscellaneous				
02h16m48s	−03°12′	Mira, Omicron Ceti, Variable Star	4	3
12h50m00s	−63°00′	Coalsack, Dark Nebula	57	10
17h37m42s	−32°11′	BM Scorpii, Variable Star	73	17
17h55m24s	−04°24′	Barnard's Star, Largest proper motion	77	19

14

Astronomical objects for small telescopes

Object 1 NGC55, Dunlop 507, galaxy
Constellation Sculptor
Position RA 00h12m30s, Dec. −39°50′
Chart number 1

Being less than 10 million light years distant, NGC55, a member of the Sculptor group of galaxies, is the nearest galaxy beyond our own local group.

NGC55 is an edge-on spiral that can be seen in the telescope as an 8th magnitude, elongated narrow streak measuring 25 by 3 arc minutes.

As with all galaxies, NGC55 is best viewed away from city lights, where it is readily visible in binoculars. An 80 to 100mm telescope will provide an impressive view. The galaxy has a definite brightening near the centre, with one edge remaining bright due to a prominent dust cloud on this side, whilst the other side fades towards the extremities. Telescopes in the 250 to 300mm range will show a mottled appearance. Large, professional instruments can resolve stars in this close neighbour.

Object 2 NGC104, 47 Tucanae, globular
Constellation Tucana
Position RA 00h21m54s, Dec. −72°21′
Chart number 2

47 Tucanae is generally rated as the second finest globular in the entire heavens, first place going to Omega Centauri. I, however, rank this as my favorite globular as it displays a delicate, fine appearance that is lacking in the brighter star city, with its coarse texture.

At 4.5 magnitude, it is easily seen with the unaided eye, as a fuzzy patch just over 2 degrees west of the Small Magellanic Cloud. The distance is 13 000 light-years, a little closer than Omega Centauri. Interestingly the age of 47 Tuc, estimated at around 12 000 million years, is about half that of Omega.

Although a 100mm telescope is required to begin resolution, smaller instruments will show a splendid sight. The view through instruments in the 200 to 300mm range is truly astonishing. Many of the Astronomical Societies listed in the back of the book have 'Star Nights' and welcome visitors. This may give you the opportunity to see this wonder through a large instrument.

Object 3 NGC253, galaxy
Constellation Sculptor
Position RA 00h45m06s, Dec. −25°34′
Chart number 1

NGC253 is one of the most prominent of the southern galaxies. It is the brightest member of the Sculptor group of galaxies; most of the others, with the exception of NGC55, are disappointing targets for the small instrument.

NGC253 is a spiral, inclined about 15 degrees to horizontal. The galaxy is framed by three stars and set in a fine field. Small instruments do well with this one, being 8th magnitude and a large 20 by 6 arc minutes. It appears as a flattened oval with central brightening, gradually fading towards the boundaries.

With a diameter of 40 000 light-years, NGC253 is not a large galaxy, at least not compared with our own, which is thought to be in the order of 100 000 light-years in diameter.

93 OBJECT 1: NGC55, an edge-on spiral galaxy in Sculptor. This photo was taken with a 250mm telescope using a 35 minute exposure. PHOTO: JOE CAUCHI.

Object 4 Omicron Ceti, Mira, variable star
Constellation Cetus
Position RA 02h16m48s, Dec. −03°12′
Chart number 3

Omicron or Mira—Latin for 'Wonderful'—is a long period, pulsating type variable star. Mira was first discovered in 1596 by Dutchman David Fabricus; he believed the star to be a nova. In 1603 it was observed at 4th magnitude by German astronomer Johann Bayer who designated the star in his famous Uranometria, the first complete star atlas, as Omicron Ceti—the 15th brightest star in the constellation Cetus.

By 1660 it was established that Omicron was a periodic variable, and it has since become the standard for this class of star. Characterised as red giants with 100 times the radius of our Sun, but with comparable mass, they have periods in excess of 100 days, with large fluctuations in brightness—from 5 to 9 magnitudes. Mira stars are very common throughout the galaxy.

With an amplitude ranging from 2nd to 10th magnitude, and a period of approximately 331 days, Mira makes an interesting subject of study for the small telescope. With a little practice at estimating magnitudes, using nearby stars of known magnitude, a graph can be drawn which will dramatically illustrate the light curve of this truly 'Wonderful' star. Its distance from our Solar System is 800 light-years.

Object 5 Theta Eridani, double star
Constellation Eridanus
Position RA 02h56m24s, Dec. −40°30′
Chart number 4

Considered by E.J. Hartung (1968) as one of the gems of the southern sky, this binary system is magnificent in small apertures. With a respectable 8.5 arc second separation, the two suns, at magnitudes 3.4 and 4.4 make a fine sight in a sparse field. They seem matched in colour, but observers tend to differ on the exact hue. Some see them as yellowish white, others as white. I see them as bluish white in an 80mm refractor; what do you see?

Object 6 M45, the Pleiades, the Seven Sisters, open cluster
Constellation Taurus
Position RA, 03h43m54s, Dec. +23°58′
Chart number 5, 5c

Even to the unaided eye the Pleiades form a striking group. In fact the cluster, immersed in a cloud of dust, illuminated by its blue giant stars, is a staggering 100 minutes of arc wide. Because of its large size, binoculars are all that are required to appreciate the full splendour that is M45. Over 100 stars are visible with a medium sized telescope, but the view is less striking.

Seven stars are visible to most people with the unaided eye, hence the name Seven Sisters. Under ideal conditions, keen eyed observers have noted up to twelve. As a test on how many you can see, try drawing the Pleiades, and compare your sketch with the chart. Make sure you check the chart last, so that you will not have a preconceived idea of each star's location.

The brightest star in the group is the 3rd magnitude Eta Tauri, or Alcyone. In decreasing order of brightness, from 3.5 to 5.5 magnitudes, the other prinicipal members are Atlas, Electra, Maia, Merope, Taygeta, Pleione, Celaeno and Asterope.

Wispy swirls of blue reflection nebulosity enhance colour photographs of the region, each bright star seemingly encased in its own little envelope.

94 OBJECT 2: NGC104, one of the finest globular clusters in the sky, is second only to ω Centauri. A 12 minute exposure with a 250mm telescope. *Photo: Joe Cauchi*

Under ideal dark sky conditions, a 100mm telescope will allow you to glimpse some of the nebulosity, particularly the brighter oval portion known as the Merope nebula, surrounding the stars of that name.

Tennyson was inspired to write:

> Many a night I saw the Pleiads
> Rising through the mellow shade
> Glitter like a swarm of fireflies
> Tangled in a silver braid.

There are several easy, wide pairs in the cluster, Atlas and Pleione for example can be separated in binoculars. With a distance of 5 arc minutes, it is hardly a test for any optical instrument.

Object 7 NGC1535, planetary nebula
Constellation Eridanus
Position RA 04h12m06s, Dec. −12°52′
Chart number 5

This much underrated planetary nebula seems to be forgotten by many modern day observers. Abbe Nicolas-Louis de Lacaille described it as the most interesting and extraordinary object of the kind he had ever seen. Herschel was certain that it was a 'very distant and highly compressed globular star cluster'.

At a distance of 2100 light-years, the pale bluish disc of NGC1535, set in a starry field, is 9th magnitude and measures 20 by 17 arc seconds. Use a low power eyepiece, as it will not bear magnification well. The central star at 11.8 magnitude requires a moderate sized instrument.

Object 8 Herschel 3670, double star
Constellation Reticulum
Position RA 04h33m00s, Dec. −62°56′
Chart number 2

A superb binary with colour contrast. The spectral types indicate pale orange and yellow, but strangely most observers see them as yellow and blue; what do you see?

The duo presents an easy object for the smallest aperture. The primary at 5.9 magnitude is separated from its 8.4 magnitude companion by 32 seconds of arc.

Object 9 Iota Pictoris, Dunlop 18, double star
Constellation Pictor
Position RA04h49m48s, Dec. −51°48′
Chart number 4

A pretty double set in a fine field. The two suns, 5.6 and 6.4 magnitude, are separated by 12 arc seconds, making them an easy target for small apertures. Both stars appear an elegant yellow, which agrees with their spectral class of FO; however, James Dunlop in 1826 described them as white and blue.

Object 10 Beta Orionis, Rigel, double star
Constellation Orion
Position RA 05h12m06s, Dec. −08°15′
Chart number 5

A superb double separated by a respectable 9.5 arc seconds. The 6.7 magnitude companion to the 0 magnitude bluish/white supergiant can be a difficult object for the novice. Easily visible in a 60mm refractor at 50 power, the overpowering brilliance of Rigel tends to flood the field with light, making detection of the companion difficult.

To overcome Rigel's glare, try observing in twilight when you can just pick out Orion's belt stars with the unaided eye.

Rigel, some 900 light-years distant, is one of the most luminous stars known in our galaxy, and the seventh brightest star in our sky.

Object 11 Large Magellanic Cloud, LMC, Nebecula Major
12 Small Magellanic Cloud, SMC, Nebecula Minor
Constellation Dorado and Mensa (LMC) Tucana (SMC)
Position RA 05h20m00s, Dec. −69°00′ (LMC)
RA 00h50m00s, Dec. −73°30′ (SMC)
Chart number 2

Dominating the southern-most part of the sky, are two large hazy nebulous regions known as the Large and Small Magellanic Clouds. They were named in honour of the Portuguese explorer Magellan, after his circumnavigation of the world in 1518-20. Both clouds are circumpolar from mid-southern latitudes.

The extragalactic duo, members of the Local Group, are thought to be satellite galaxies of our own, orbiting with a period measured in many millions of years. The Local Group of Galaxies includes about 30 galaxies within a 500 000 light-year sphere. Most of the components of this group tend to congregate around the two brightest members, the Andromeda galaxy and our own galaxy. The majority of these galaxies are insignificant dwarfs. With the exception of a faint dwarf galaxy somewhat closer, the Clouds of Magellan are the nearest of the external galaxies.

The distance of the LMC is 160 000 light-years, and the SMC a little further at 200 000 light-years. The angular distance separating the clouds is about 20 degrees, corresponding to an actual distance of 80 000 light-years between the two. It is indeed fortuitous to have two galaxies at our back doorstep, providing astronomers with a unique laboratory to study these 'Island Universes' in far greater detail than any other galaxy, apart from our own. It is interesting to note that a 100mm telescope will show as much detail in the clouds as 1000mm of aperture can on the bright and relatively close Andromeda galaxy.

Sprawling across 6 degrees of sky, the LMC is classified midway between an irregular and a barred spiral-type galaxy, and abounds with globulars, nebulae, and clusters. The SMC, an irregular galaxy, is smaller than its neighbour, covering 3.5 degrees. It tends to be overshadowed by the LMC, but, on a reduced scale, it too has many objects of interest.

Australian astronomers have discovered that the SMC is actually two galaxies, one superimposed on the other, the far galaxy being about 30 000 light-years beyond the SMC. This has led to speculation by astronomers that a collision occurred between the LMC and the SMC in the

95 OBJECT 3: NGC253, another of the spiral galaxies in Sculptor. A 30 minute exposure with a 200mm telescope. PHOTO: KEN WALLACE

96 OBJECT 6: M45, the Pleiades, a 15 minute exposure using a 250mm reflector. PHOTO: JOE CAUCHI

distant past, the encounter tearing the SMC in two.

Both galaxies are easily visible to the unaided eye, resembling detached portions of the Milky Way. The sight under very dark country skies leaves a lasting impression; even the presence of the moon fails to extinguish their soft glow. Binocular owners will find a wealth of detail, and a small telescope, at low magnification, allowed to drift slowly across each, will unfold a panorama with few celestial equals. The Tarantula nebula, NGC2070, so prominent a part of the LMC, is treated separately in these pages.

Object 13	NGC1952, M1, the Crab nebula, planetary nebula
Constellation	Taurus
Position	RA 05h31m30s, Dec +21°59′
Chart number	5

Charles Messier accidentally stumbled across the Crab nebula in 1758 while observing a comet. This prompted him to list other nebulous objects so astronomers would not confuse them with comets. And so the famous Catalogue of Messier was born, with the Crab nebula as object number one. John Bevis, an English amateur, discovered M1 27 years before Messier in 1731.

Located about 70 arc seconds northwest from the 3rd magnitude star Zeta Tauri, the 8th magnitude Crab may be detected in binoculars or finder scopes. An 80mm telescope will show the 6 by 4 arc minute oval shape well. Larger instruments, 250 to 300mm, are required to see detail in the serrated extremities.

It is interesting to note that you are observing the remnant of a supernova explosion that occurred on 4 July 1054. The supernova was so bright that Chinese astronomers were able to see it for 23 days in broad daylight! After more than 12 months the 'guest star' faded beyond naked-eye visibility. The expanding cloud of gas from the explosion is all that the amateur sees today. A faint 16th magnitude pulsar is all that is left of the original star. The distance of M1 is 6000 light-years, and the diameter is about 10 by 7 light-years.

Object 14	NGC1976, M42, the Great nebula
15	Theta Orionis, the Trapezium
16	NGC1982, M43, nebula
17	42 Orionis, double star
18	NGC1977, nebula
19	NGC1981, open star cluster
20	Iota Orionis, double star
21	Struve 747, double star
Constellation	Orion
Position	RA 05h32m54s, Dec. −05°25′
Chart number	5, 5a

M42, the Great nebula in Orion, offers a spectacle with few equals. Possibly the most photographed and studied objects outside our solar system, by both professional and amateur astronomers.

This vast complex cloud of swirling ionised hydrogen gas, illuminated by the young hot stars within its central regions, is but the chaotic heart of a sea of fainter nebulosities that extend throughout the entire constellation of Orion.

At a distance of about 1300 to 1500 light-years, the main portion of M42 visible in small telescopes is about 30 light-years across. Within its depths new stars are forming.

Immersed in this celestial birthplace is Theta Orionis, a young quadruple star system known as the Trapezium. These young hot stars which formed as recently as 100 000 years ago, are in part responsible for the view of M42 that we see today. The obscuring material has gradually been pushed away from the hot, infant stars, like an enormous bubble. On our side of the nebula the bubble burst, and we now look into the cavernous interior of this gigantic dust cloud. The Trapezium stars are part of a cluster, most of whose members are still forming and are hidden deep behind obscuring matter. Still not hot enough to blow away their dusty envelopes, these

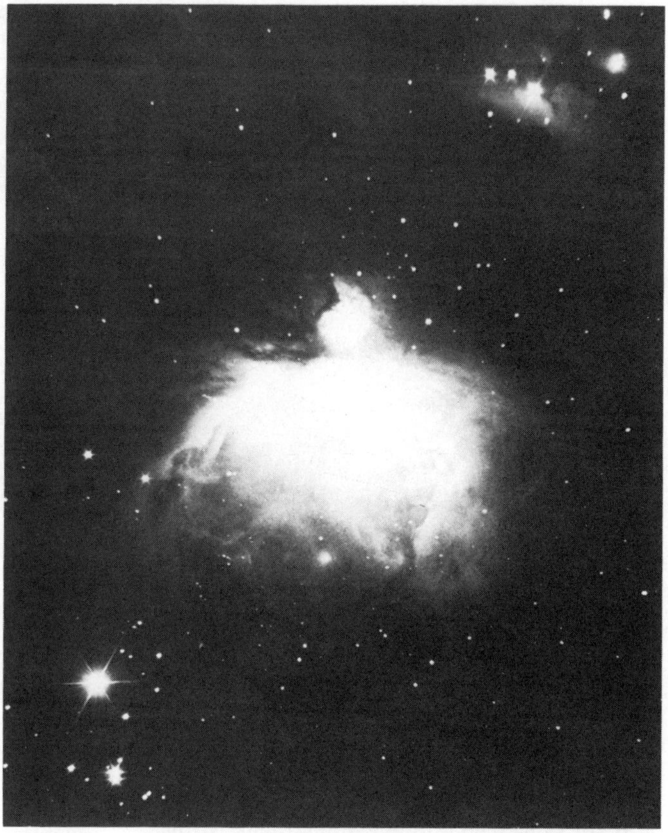

97 OBJECT 14: NGC1976, the Orion nebula. This 12 minute exposure was made using a 250mm telescope. Because of the length of the exposure needed to show the nebulosity you cannot see the stars of the trapezium within the gas cloud. PHOTO: JOE CAUCHI

stars are visible only with penetrating infrared photography.

Even for the city dweller under light polluted skies, M42 is easily visible, having the highest surface brightness of any nebula. It is easily located surrounding Theta Orionis, the middle of the three stars which form the sword of Orion. Under good, dark skies, a keen eye will perceive Theta Orionis as ill-defined and hazy in appearance. The binocular view is stunningly beautiful, and the low power field in a small telescope reveals bright mottled condensations, filamentary streamers and dark intrusions. From the bright centre, two luminous wings arc away, nearly joining to form a ghostly, nebulous wreath.

Observers frequently report a grey/green tint to M42, which is in direct contrast to the red coloration in photographs. This discrepancy is due to the fact that colour photographs discern the red hydrogen light, and the human eye, being more sensitive to the middle of the spectrum, detects the green, oxygen emissions. All other nebulae, being fainter, are not sufficiently bright to activate the colour sensitive parts of the retina.

Easily visible in a small instrument is the famed Trapezium; by contrast it appears to lie in a dark region of the nebula—an 80mm refractor at 30 power is sufficient to see the four stars. More difficult however are the 5th and 6th members, designated as stars E and F in chart 5b. These two stars at 11th magnitude can be a severe test for the small instrument, but may under excellent conditions and high magnification be observed.

Like the Eta Carinae nebula, the Great nebula in Orion requires careful scrutiny to appreciate all the intricate features.

In the immediate vicinity of the Great nebula are several other interesting objects. In the same field and to the north of Theta Orionis lies M43, appearing like a detached portion of M42 in the telescope but joined in photographs. Half a degree to the north is the 4.6 magnitude star 42

98 OBJECT 22: NGC2070, the Tarantula nebula, is within the Large Magellanic Cloud, a small companion galaxy of our own. This 30 minute exposure shows a wide view of the area taken with a 125mm Schmidt camera. PHOTO: KEN WALLACE

Orionis which is immersed in a small haze of nebulosity known as NGC1977. A little further north is the scattered open cluster NGC1981 which contains about a dozen stars of magnitudes 8 down to 10.

Half a degree to the south of M42 is Iota Orionis, an attractive triple. The components of this system are magnitudes 2.9, 7.4, and 11, separated by 11.5 and 50 arc seconds respectively. In the same field is found another pair, Struve 747, a dainty white pair of magnitudes 5.6 and 6.7 separated by 36 seconds of arc. The whole star-studded field of Iota Orionis is a real treat.

Object 22	NGC2070, Dunlop 142, Tarantula nebula
Constellation	Dorado
Position	RA 05h39m54s, Dec. −69°04′
Chart number	2

Located within the Large Magellanic Cloud is the spectacular Tarantula nebula, surrounding the star 30 Doradus. At the LMC's distance of 160 000 light-years, it is the most distant of the gaseous nebula to be seen with the unaided eye, and the only nebula within another galaxy visible in a small instrument.

Binoculars afford an excellent view, indeed the whole LMC lends itself to this type of optical aid. Small telescope owners will be rewarded with views of bright luminous areas, contrasted against dark lanes. Instruments in the 200 to 300mm range and country skies, reveal the extremely complex looped structure that is responsible for the names 'Tarantula' and 'Great Looped' nebula.

The diameter of the Tarantula is about 900 light-years, making it the largest observable gaseous nebula in the entire universe! If the Great nebula in Orion, M42, were to be replaced by the Tarantula, it would extend a staggering 35 degrees across the sky—seventy moon diameters! It would glow with an integrated brightness sufficient to cast shadows. Sir John Herschel's 1830s description is fitting: 'This is one of the most singular and extraordinary objects which the heavens present . . .'

Object 23	NGC2168, M35, open star cluster
24	NGC2158, open star cluster
Constellation	Gemini
Position	RA 06h05m42s, Dec. +24°20′
Chart number	5

Under good dark skies, this 5.5 magnitude open star cluster is just visible to the unaided eye, and is excellent in binoculars. M35 is comprised of about 120 stars ranging in magnitude from around 9th to 16th magnitude; these are spread evenly within its 30 arc-minute boundary. The cluster is splendid in small apertures: with low magnification the stars form arcs and lines. A pretty orange 8.6 magnitude sun lies near the

99 OBJECT 23: NGC2168. This open star cluster in Gemini is shown in a 10 minute exposure through a 250mm reflector. PHOTO: JOE CAUCHI

centre. M35's distance is about 2800 light-years.

If you feel like a challenge, and have around 100mm of aperture, try for the compact 4 arc-minute diameter, 12th magnitude cluster NGC2158, lying just half a degree to the south-west, but at a distance of 15 000 light-years. The exact nature of NGC2158 is uncertain—it resembles a globular, but lacks the stellar population normally associated with them.

Object 25 Beta Monocerotis, multiple star
Constellation Monoceros
Position RA 06h26m24s, Dec. −07°00′
Chart number 7

Described by Sir William Herschel in 1781 as 'one of the most beautiful sights in the heavens', this triple star system forms a narrow triangle. The three stars have similar spectra and appear brilliant white. The main pair, 4.7 and 4.6 magnitude, are separated by 7.4 arc seconds; a close 2.8 arc seconds from the 4.6 magnitude star is another of magnitude 5.6. This triple is a good test for a small telescope—first use a high enough magnification to split the main pair, and if the seeing is good, try a higher power to separate the close duo. You will be rewarded with a rare sight, as few triples have all components of almost equal magnitude. I find under good conditions, an 80mm refractor at 80 power is sufficient to show all three effortlessly.

Object 26 Dunlop 30, Rumker 65, double star
Constellation Puppis
Position RA 06h28m36s, Dec. −50°12′
Chart number 6

A fine pair of stars, the primary at 5.3 magnitude appears yellow and the secondary is distinctly reddish at 9th magnitude. Separation of the suns is 12.4 arc seconds. Interestingly both stars have very close companions of almost equal magnitude; unfortunately steady seeing and a large amateur instrument, at high magnification, is required to separate them.

Object 27 NGC2287, M41, open star cluster
Constellation Canis Major
Position RA 06h44m54s, Dec. −20°42′
Chart number 7

Located 4 degrees from Sirius, the brightest star in the sky, this scattered cluster is visible to the unaided eye as a 5th magnitude hazy star. Binoculars will resolve some of the brighter members, and a small telescope at low power will reveal the full beauty of the cluster, with many of its stars forming arcs.

A conspicuous reddish star near the centre of the loose half degree field makes a striking contrast with the many blue giants. The distance of the cluster is about 2300 light-years.

Object 28 Epsilon Canis Majoris, double star
Constellation Canis Major
Position RA 06h56m42s, Dec. −28°54′
Chart number 7

A pretty double that can be a difficult test for small apertures, due to the brightness of the primary. In a fine field, the white 1.5 magnitude star has an 8th magnitude yellowish companion sun situated 8 seconds of arc away.

Object 29 NGC2362, open star cluster
Constellation Canis Major
Position RA 07h16m36s, Dec. −24°52′
Chart number 7

A magnificent cluster surrounding the 4th magnitude star Tau Canis Majoris. Measuring only 6 arc minutes across, even the smallest of instru-

100 OBJECT 29: NGC2362, an open star cluster in Canis Major. This picture was made using a 250mm telescope and a 12 minute exposure. PHOTO: JOE CAUCHI

ments will reveal about 20 or so stars. Tau itself is a spectroscopic binary, with two other companion suns. One lies 8 arc seconds away at 10th magnitude, and the other at 14 arc seconds at 11th magnitude. Less than 1 000 000 years old, NGC2362 is one of the youngest known open clusters, and is 4500 light-years distant.

Object 30 Struve 1097, double star
Constellation Canis Major
Position RA 07h25m30s, Dec. −11°27′
Chart number 7

A splendid multiple star. The 5.5 magnitude yellowish primary has a close companion just 0.7 seconds of arc away and requires a large aperture to separate it. Two other white suns, 8.5 and 9th magnitude, are easy for small instruments and are located 20 and 23 seconds of arc away.

Object 31 NGC2437, M46, open star cluster
32 NGC2422, M47, open star cluster
33 NGC2438, planetary nebula
34 Struve 1120, double star
35 Struve 1121, double star
Constellation Puppis
Position RA 07h39m18s, Dec. −14°42′
Chart number 7

M46 is a rich open star cluster, appearing circular with no real condensation towards the centre. The stars seem fairly evenly distributed and tend to form arcs and lines. At 25 arc minutes in diameter with an integrated magnitude of 6, it is easily detected in binoculars, together with the brighter M47 (see below). Several hundred stars are known to be members of the cluster; the brightest 50 or so with magnitudes 10 and fainter are seen with a small instrument at low magnification.

A small annular planetary nebula, NGC2438, lies to the north of centre. It is visible in instruments with at least 100mm of aperture, although good dark skies are a must, as it is only 11th magnitude. This planetary is not a true member of M46 but is situated between us and the cluster. The distance to M46 is about 5000 light-years and the diameter is in the order of 30 light-years.

Just 1.5 degrees to the west is another open cluster, M47, which is in direct contrast to M46. Within its 20 arc-minute diameter 25 stars of magnitude 6 and fainter congregate, forming an attractive uneven scattering of bright and fainter suns. Two orange 8th magnitude stars make a nice contrast among the predominantly blue and white population. Two double stars are associated with the cluster, Struve 1120 and Struve 1121. Whilst you should view the cluster at very low magnification, a moderate power will assist in locating both doubles. Struve 1121 is a very neat pair, with both components of equal 7.9 magnitude, and separated by 7.5 arc seconds. Struve 1120 is seen as stars of 5.6 and 9.5, separated by 20 arc seconds. Both doubles are

101 OBJECT 31: NGC2437, an open star cluster in Puppis. The cluster, also called M46, has a small planetary nebula in the lower right corner. The photo was taken with a 250mm reflector and a 25 minute exposure. PHOTO: JOE CAUCHI

actually triples but the third companion to each requires a large amateur instrument to detect.

As would be expected the bright, 30 light-year diameter M47 is closer than M46, the estimate being about 1700 light-years.

Object 36 k Puppis, double star
Constellation Puppis
Position RA 07h36m48s, Dec. −26°41′
Chart number 7

An excellent double set in a fine starry field. The pair is separated by 9.8 seconds of arc, and appear almost equally matched, being 4.5 and 4.6 magnitude. Both stars appear to be pale yellow, despite their spectra indicating a bluish colour. A third component of the 14th magnitude is beyond the range of small instruments. A 75mm refractor at 100 power gives a pleasant view.

Object 37 NGC2477, open star cluster
Constellation Puppis
Position RA 07h50m30s, Dec. −38°25′
Chart number 6

With over forty open star clusters abounding in the constellation Puppis, it is difficult to decide on a favourite, but NGC2477 would rank high on my list. It is easily located by moving about 2.5 degrees north-west of Zeta Puppis, a 2.25 magnitude blue-white supergiant, to a 4th magnitude star; NGC2477 is immediately north of this star.

With an integrated magnitude of 6, NGC2477 presents an easy binocular target; the field is excellent. Set in a rich region of the Milky Way, the view within a 5 degree radius of Zeta Puppis is very rewarding. Moving in a north-west line from Zeta you see NGC2477 followed by NGC2451, a large scattered grouping of bright stars. North-east of Zeta is NGC2546, a more compressed open cluster.

In the telescope, NGC2477 is about 25 arc minutes across, and condensed towards the centre. With about half the population of 300 11th magnitude stars congregated within a 12 arc-minute ball, the appearance is that of a loose globular cluster. A great many stars form interesting patterns, spiralling arms and lines are easy to detect.

A good dark sky and at least 100mm of aperture are required to bring out the best in this object.

Object 38 NGC2516, open star cluster
Constellation Carina
Position RA 07h59m42s, Dec. −60°44′
Chart number 8

A most impressive open star cluster with more than 100 stars strewn over a one degree diameter field. At magnitude 3.8 it is easily visible to the

102 OBJECT 32: NGC2422, another open cluster in Puppis. This time the photo is a 10 minute exposure with a 250mm reflector. PHOTO: JOE CAUCHI

unaided eye and makes a fine binocular subject. A few gleaming orange stars and the contrast between the other bright and a faint members make this cluster one of the best. As a bonus there are three pretty doubles that are easily resolvable in small instruments. NGC2516 is 80 light-years in diameter, and 4300 light-years distant.

Object 39 Gamma Velorum, Dunlop 65, multiple star
Constellation Vela
Position RA 08h07m54s, Dec. −47°12′
Chart number 6

A beautiful mutliple star system, the main pair easily divisible in binoculars. The primary, a very hot white luminous giant, of magnitude 2.2, makes a fine sight with its bluish white, 4.8th magnitude companion, at a very wide 41 arc seconds. There are two more components almost at right angles to the main pair, at magnitude 8.5 and 9.4, separated by 40 arc seconds; both appear blue. Set in a star-studded field, this is an excellent object for small apertures.

Object 40 NGC2632, M44, the Beehive, open star cluster
Constellation Cancer
Position RA 08h37m30s, Dec. +19°52′
Chart number 9

At a distance of 450 light years, the Beehive or Praesepe is one of the nearest open star clusters. Easily visible to the unaided eye as a hazy one degree diameter fuzzy patch, it only requires binoculars to resolve the cluster into a myriad of stars.

Fifteen of the brightest members range in magnitude from 6.5 to 7.5. Telescopes using low magnification will reveal several double and multiple stars. Galileo was the first to discover that the Beehive was in fact composed of stars, when he turned his tiny instrument towards the cluster in 1610.

Object 41 NGC3132, the Eight-Burst nebula
Constellation Vela
Position RA 10h04m54s, Dec. −40°11′
Chart number 11

Located on the border of the constellations Vela and Antlia is the conspicuous planetary nebula known as the Eight-Burst. Several rings of nebulous material appear superimposed at different angles as if the central star has had many outbursts, hence the name. These rings only show up on long-exposure photographs, giving NGC3132 an overall size of 80 by 50 minutes of arc.

With an integrated magnitude of 8, the oval shaped nebula, in a fine field, is easily discernible in small apertures, as is the 10th magnitude central star. The observable star is not responsible for illuminating the nebula, the culprit being a faint 16th magnitude dwarf companion. Distance estimates vary; about 2000 light-years seems likely.

103 OBJECT 37: NGC2477, a third open star cluster in Puppis. An 18 minute exposure with a 250mm telescope. PHOTO: JOE CAUCHI

Object 42 Gamma Leonis, double star
Constellation Leo
Position RA 10h17m12s, Dec. +20°06′
Chart number 12

An outstanding pair, consisting of two suns of magnitudes 2.6 and 3.8, separated by 4.3 seconds of arc. Over the past two centuries observers just have not seemed able to agree on the colours of the pair. Gold and greenish, yellow and green, white and red, are some of the quoted estimates. With an 80mm refractor I see them both as deep yellow; how do your observations compare? With a pair so close and bright, some difficulty may be experienced in splitting them with small apertures; try high magnification if the seeing is good, or even observe them during twilight to reduce the glare.

Object 43 NGC3242, Ghost of Jupiter, planetary nebula
Constellation Hydra
Position RA 10h22m24s, Dec. −18°23′
Chart number 13

This bright 9th magnitude planetary nebula is known as the Ghost of Jupiter. It is easy in small instruments—with medium to high magnification it will be seen as a bluish oval-shaped nebula about 25 by 15 arc seconds. The bright oval itself is surrounded by a fainter outer envelope 40 by 35 arc seconds. The secondary shell and the 11.5 magnitude central star require at least a 250mm telescope. The estimated distance is about 2000 light-years.

Object 44 Herschel 4330, t Velorum, double star
45 Herschel 4332, double star
Constellation Vela
Position RA 10h30m48s, Dec. −46°45′
Chart number 11

The aesthetic appeal of binary stars is enhanced when more than one pair happens to occupy the same field. This occurs frequently, but the following delightful duo is one of my favourities.

h4330 consists of a superb 5.1 magnitude yellow primary, separated by 40 arc seconds from its blue 8.6 magnitude companion. Two faint, unrelated field stars point in the direction of h4332: a pair of white and blue suns, 7.2 and 9.6 magnitude, separated by 28 arc seconds. Enriched by a starry field, the view of the colourful pairs is excellent with a small instrument.

Object 46 NGC3293, the Gem Cluster
Constellation Carina
Position RA 10h31m30s, Dec. −57°58′
Chart number 10

Located just north-west of the extremities of the Eta Carinae nebula is a superb little 8 arc-minute diameter galactic cluster. Aptly named the Gem

Cluster by Charles Rumker in the 1820's, this delicate compact group features a red supergiant, surrounded by blue and blue/white suns, magnitudes 6 down, against a backdrop of fainter stars. Excellent in any telescope from 60mm up, the cluster, with about 50 members, is compressed over 3.5 light-years and lies at a distance of 1400 light-years.

Object 47 I.2602, open star cluster
48 Mel 101, open star cluster
Constellation Carina
Position RA 10h41m00s, Dec. −64°08′
Chart number 10

I.2602 is a bright open cluster surrounding the 2.8 magnitude star Theta Carinae. Several members of the group are visible to the naked eye, and the view through binoculars reveals the reason for the nickname, the Southern Pleiades. Spread out over 70 arc minutes, this striking collection of 30 blue/white suns, is superb in any small instrument. With a distance of 650 light-years it is one of the closest clusters.

Within 30 arc minutes south of I.2602 is another cluster, Mel 101. At a distance of 7500 light-years, its 40 faint stars make a difficult subject. To detect this dim 15 arc-minute diameter cluster, you will require at least 80mm of aperture and dark skies.

Object 49 NGC3372, Dunlop 309, the Eta Carinae nebula
Constellation Carina
Position RA 10h43m06s, Dec. −59°25′
Chart number 10

The Eta Carinae nebula is one of the largest and brightest of the nebulae; it is visible to the unaided eye as a bright region in a rich area of the southern Milky Way. Binoculars reveal several wedge-shaped nebulous sections, covering several square degrees, separated by opaque dust lanes. Although the nebula is best appreciated in its entirety with binoculars, try sweeping your telescope at low magnification through the region. Double stars, clusters, dark and light nebulae all combine to make this an area of beauty and complexity with few equals.

Sir John Herschel best described the nebula in the 1830's with the following: 'Nor is it easy for language to convey a full impression of the beauty and sublimity of the spectacle it offers when viewed in a sweep, ushered in as it is by so glorious and innumerable a procession of stars, to which it forms a sort of climax, justifying expressions which, though I find them written in my journal in the excitement of the moment, would be thought extravagant if transferred to these pages'.

Embedded within the most luminous section of nebulosity is the remarkable orange star Eta Carinae; Eta in turn is enveloped in a small nebula known as the Homunculus. Recorded as 4th magnitude in 1677, it has undergone some

104 OBJECT 38: NGC2516, an open star cluster in Carina. At magnitude 3.8 this object can easily be seen with the naked eye. A 15 minute exposure with a 250mm telescope. PHOTO: JOE CAUCHI

spectacular variations in brightness and was, in 1843, the second brightest star in the sky at −0.8 magnitude. With comparatively minor fluctuations over the past 100 years, it has remained at 7th magnitude. Recent observations indicate a gradual brightening, and some astronomers predict that Eta will reach 3rd magnitude by the year 2000. With Eta Carinae being one of the main stars illuminating the nebula we may find the sight even more majestic in years to come. Due to the tiny Homunculus nebula, Eta never looks quite focused in a telescope.

Near Eta Carinae is the Keyhole nebula (many people incorrectly give this name to the entire Eta Carinae nebula), a small dark region described by Herschel as a 'singular lemniscate-oval vacuity which forms so strange a feature of this nebula'. Herschel's drawings at the time clearly show an old fashioned keyhole shape. Since fading, Eta no longer illuminates the near side of the keyhole and the original shape is now lost; however, a small dark bay can still be observed.

Object 50	NGC3532, Dunlop 323, open star cluster
Constellation	Carina
Position	RA 11h03m24s, Dec. −58°24′
Chart number	10

Rated by many observers as the finest open cluster in the heavens. Visible to the unaided eye at magnitude 3 and situated within 3 degrees of the Eta Carinae nebula. Up to 150 stars of magnitude 8 to 12 are located within a 60 by 30 arc-minute area, giving the appearance of a football with a bright yellow star at one end. Magnificent in any instrument. The distance to the cluster is about 1300 light-years and its diameter is 25 light-years; in total over 400 stars are known to be members.

Object 51	NGC2623, M65, galaxy
52	NGC3627, M66, galaxy
53	NGC3628, galaxy
Constellation	Leo
Position	RA 11h16m18s, Dec. +13°23′
Chart number	12

An interesting trio of galaxies for the small telescope user. Both M65 and M66 are visible in good binoculars. At 35 million light-years distant they appear to be outer members of the giant Virgo Cluster of galaxies.

M65 is nearly edge-on at 9.3 magnitude, measuring 8 by 1.5 arc minutes. In the same field at a close 20 arc minutes lies M66 at 8.4 magnitude; not as edge-on as M65, it measures 8 by 2.5 arc minutes. Although M66 is the brighter of the two, M65 is more prominent due to its smaller surface area. Using an instrument of 100mm aperture M65 will be seen as a bright silver spindle with the nearby M66 displaying a mottled appearance with some dark regions.

In 1773 Charles Messier observed a comet passing between the two galaxies, but failed to notice them due to the brightness of the comet. Their discovery by Messier was to come some 7 years later.

105 OBJECT 40: NGC2632, Praesepe in Cancer, is one of the nearest open star clusters to Earth. This 12 minute exposure with a 250mm telescope shows the individual stars in the group quite well. PHOTO: JOE CAUCHI

106 OBJECTS 47 AND 48: I.2602 and Mel 101 are two bright open star clusters in Carina. I.2602 at lower left is 650 light-years from Earth while Mel 101 at right is 7500 light-years from us. This 15 minute exposure was made using a 125mm Schmidt camera. PHOTO: KEN WALLACE

Frequently missed by observers is the third member of the trio, NGC3628. Lying just 35 arc minutes north of Messier's pair, it is larger than the others at 12 by 1.5 arc minutes, but fainter with a total magnitude of 10.9. In a small telescope you will see the elongated brighter northern half of the galaxy; a dark dust lane obscures much of the southern side.

Object 54 NGC3766, Dunlop 289, open star cluster
Constellation Centaurus
Position RA 11h34m00s, Dec. −61°19′
Chart number 10

This 15 arc-minute diameter open cluster is located in a rich field midway between the Eta Carinae nebula and Alpha Crucis, the brightest star in the Southern Cross. A population of about 50 stars is visible in small instruments, ranging from 8th magnitude down. A few scattered bright stars can be seen to the north and make an attractive low power field. It is excellent in binoculars or finder scope. NGC3766 is 10 light-years in diameter, and 1700 light-years distant.

Object 55 NGC3918, planetary nebula
Constellation Centaurus
Position RA 11h47m48s, Dec. −56°54′
Chart number 10

This bright 8th magnitude planetary nebula is easily visible in the smallest of instruments, and appears starlike. By increasing the magnification you will see a pale blue/green disk. This object is a prime example of how the term 'planetary nebula' originated. Set in a rich star field this little gem looks similar to the planet Uranus. The distance has been estimated at 14 000 light-years.

Object 56 Acrux, Alpha Crucis, multiple star
Constellation Crux
Position RA 12h23m48s, Dec. −62°49′
Chart number 10

Acrux or Alpha Crucis, at 1st magnitude, is the brightest star in the Southern Cross, and one of the finest triple star systems in the heavens. The wide bluish pair, magnitudes 1.5 and 5, are separated by 90 seconds of arc, making them an easy object for small telescopes. The close pair can be difficult and requires very steady seeing. After examining the wide pair, see if the brighter component appears elongated; if so, try to separate them using a high magnification. If you succeed you will find two bluish gems of magnitude 1.4 and 1.9, separated by 4.4 seconds of arc. An 80mm refractor should have no difficulty here.

Object 57 The Coalsack, dark nebula
Constellation Crux
Position RA 12h50m00s, Dec. −63°00′
Chart number 10

A large 7 by 5 degree dark region east of Alpha Crucis; it is very conspicuous in country skies. Best viewed with the naked eye, the Coalsack is a cloud of obscuring matter situated about 500 light-years away, effectively blocking the light from the starclouds behind. It is not however completely barren; in binoculars, many stars can be seen to shine through. A 6.5 magnitude star just visible to the unaided eye near the centre, is a very good test for magnitude limit and seeing conditions.

107 OBJECT 49: The η Carinae nebula is one of the largest and brightest of nebulae. Seen here in a 30 minute exposure with a 125mm Schmidt camera many double stars, dark and light nebulae are also shown.
PHOTO: KEN WALLACE

Object 58 NGC4755, Dunlop 301, the Jewel Box, open cluster
Constellation Crux
Position RA 12h50m36s, Dec. −60°05′
Chart number 10

A beautiful cluster located near the second brightest star in the Southern Cross, Beta Crucis. The cluster surrounds the fine red star Kappa. The six brightest stars form an 'A' shape and provide interesting contrast with the fainter members; about 50 stars, 6th to 10th magnitude, are spread over 10 minutes of arc. Sir John Herschel described the cluster in the 1830's as a 'superb piece of fancy jewellery'. Observers using large aperture instruments have described many colours within the group, but the small telescope will show mainly white or bluish stars, with the exception of the red supergiant Kappa. The Jewel Box is one of the youngest clusters known, and lies at a distance of 7800 light-years.

Object 59 Mu Crucis, Dunlop 126, double star
Constellation Crux
Position RA 12h51m36s, Dec. −56°54′
Chart number 10

This 4th magnitude pair makes a very attractive sight. The two white suns shine at magnitude 4.3 and 5.5, and are separated by 35 seconds of arc, making this an easy pair for apertures of just 60mm.

Object 60 NGC4945, Dunlop 411, galaxy
Constellation Centaurus
Position RA 13h02m24s, Dec. −49°13′
Chart number 14

Situated near Xi Centauri is the 9th magnitude nearly edge-on spiral galaxy NGC4945. An 80mm telescope will show a faint silver streak 10 minutes of arc long, situated in a good, star-studded field. NGC4945 is one of several galaxies forming the nearby Centaurus group; other members are NGC5128 and NGC5236.

Object 61 NGC5024, M53, globular cluster
Constellation Coma Berenices
Position RA 13h10m30s, Dec. +18°26′
Chart number 15

This lovely globular cluster was first observed by Bode in 1775, and independently discovered by Messier in 1777. Slightly oval in shape, with a bright compact centre, it is an easy object for the small instrument. A telescope of 100mm aperture at medium magnification should resolve some of the stars on the outer fringes if seeing is good. The centre will appear grainy, hinting at resolution. Sir John Herschel aptly described the globular as a 'brilliant mass of minute stars, blazing in centre'. M53's diameter is 10 minutes of arc, with an integrated magnitude of 7.5. The distance is 65 000 light-years.

Object 62 NGC5128, Dunlop 482, galaxy
Constellation Centaurus
Position RA 13h22m24s, Dec. −42°45′
Chart number 14

Situated less than 5 degrees north of Omega Centaurus, is the giant elliptical galaxy NGC5128. Measuring 10 by 8 minutes of arc, this 7th magnitude galaxy, divided into two semicircular segments by a dark dust lane of obscuring matter, is easily visible in small apertures and binoculars, provided that a good dark observing

108 OBJECT 50: NGC3532 is an open star cluster in Carina. Over 150 stars of the cluster can be seen in this 15 minute exposure taken with a 250mm telescope.
PHOTO: JOE CAUCHI

site is chosen. Since the RA of NGC5128 is close to that of Omega Centauri, locating 5128 is easy if you have an equatorial mount that is set up correctly. Simply find Omega and lock the RA axis, then swing the telescope gently northwards in declination—you will soon stumble onto 5128.

NGC5128 is also known as the strong radio source Centaurus A. It was this connection between the optical and radio that led to the identification of 5128 as an external galaxy. Up until that time it was believed to be a rather odd nebula within our own galaxy. At a distance of 16 million light-years, it is one of the most luminous and massive galaxies known.

Object 63	NGC5139, Omega Centauri, globular cluster
Constellation	Centaurus
Position	RA 13h23m48s, Dec. −47°13′
Chart number	14

The brightest of the globular clusters, this majestic object is easily visible to the unaided eye as an ill-defined magnitude 3.5 star. Binoculars reveal a hazy ball somewhat like a tailless comet, a description that fits well, since many mistook Omega for Halley's Comet during April 1986, when the famous comet passed within 5 degrees of the cluster.

A beautiful sight in any telescope, even small apertures will resolve individual stars. Definitely one of the showpieces of the southern skies. To quote Sir John Herschel: 'Truly astonishing . . . the richest and largest object of its kind in the heavens, whose stars are literally innumerable.'

An unusual feature of Omega is its slightly elliptical shape, indicating a faster rate of rotation than most globulars. Lying at a distance of 16 500 light-years, it has a diameter of 600 light-years, making it the largest of any known globular, and probably the most massive, with its million or more stars.

Object 64	NGC5236, M83, galaxy
Constellation	Hydra
Position	RA 13h34m18s, Dec. −29°37′
Chart number	14

The galaxy M83 was discovered by French astronomer Lacaille in 1752. Almost face-on, this spiral presents an easy target for small instruments. It appears as an ellipse 7 by 5 arc minutes, shining with an integrated magnitude of 8. At a distance of 10 million light-years, M83 is one of several galaxies forming the nearby Centaurus group; other members are NGC5128 and NGC4945.

Although at least a 200mm telescope is required to view the spiral structure, an 80 or 100mm instrument will show an oval shaped nebulous patch with a very bright small nucleus. One of the most impressive of the southern galaxies.

Object 65	NGC5286, Dunlop 388, globular star cluster
Constellation	Centaurus
Position	RA 13h43m00s, Dec. −51°07′
Chart number	14

You will find this neat little 4 arc-minute diameter, 9th magnitude globular a direct contrast to the magnificent Omega Centauri. Comprised of 14th magnitude suns, it appears as an unresolved misty ball, providing an attractive sight with a 6th magnitude yellow/orange field star. The star makes finding the object an easy chore.

109 OBJECTS 51, 52 AND 53: These three galaxies in Leo show different forms. M66 at the top of the photo is a spiral galaxy seen at a good angle. M65 at lower left is another spiral seen almost edge-on, while NGC3628 at right is a fainter edge-on spiral. This 28 minute exposure with a 250mm telescope shows all these galaxies to good advantage.
PHOTO: JOE CAUCHI

A small instrument will show this one well, and one could be forgiven for mistaking it for NGC6441 in Scorpius; the similarity is remarkable. NGC5286 is 35 light-years in diameter, and 38 000 light-years distant.

Object 66 Alpha Centauri, multiple star
Constellation Centaurus
Position RA 14h36m12s, Dec. −60°38′
Chart number 10

At magnitude 0.06 Alpha Centauri is the third brightest star in the sky, outshone only by Sirius and Canopus. Alpha is a multiple star system and at a distance of 4.3 light-years, is the closest binary to our solar system. The smallest of telescopes will resolve the main pair into two dazzling white headlights, of magnitude 0.33 and 1.7, with a current separation of 21 seconds of arc.

At 0.2 of a light-year from the main pair, and 4.2 light-years from our Solar System, is Proxima Centauri, a faint red dwarf flare star. Proxima has the distinction of being the closest of our stellar neighbours, apart from our own Sun. While the two bright stars orbit each other once every 80 years, Proxima in turn orbits them in a period measured in millions of years. At magnitude 11, it appears as an unimpressive object in a field strewn with other faint stars about 2 degrees south-west of Alpha.

Object 67 Alpha Circini, Dunlop 166, double star
Constellation Circinus
Position RA 14h38m24s, Dec. −64°46′
Chart number 10

This pretty double, magnitudes 3.4 and 8.8, is separated by 15 arc seconds, and is an easy

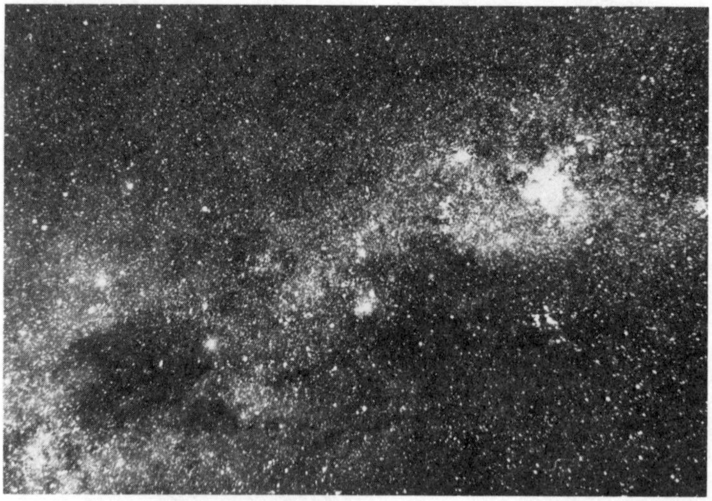

110 OBJECT 57: The Coalsack nebula is a very good example of a dark nebula, situated as it is in a bright region of the Milky Way just below the Southern Cross. See if you can find the Cross in this 10 minute piggy-back exposure taken with a 50mm lens. The η Carinae nebula can also be seen at the left of this photo. PHOTO: STEPHEN BURNS

object with an 80mm telescope. Herschel described the pair as white and yellow, however, I see the primary as yellow with the fainter companion as red. This is in fair agreement with their spectra; how do your observations compare?

Object 68	NGC5904, M5, globular
69	5 Serpentis, Struve 1930, double star
Constellation	Serpens
Position	RA 15h16m00s, Dec. +02°16′
Chart number	16

Like many objects in Messier's catalogue, M5 was discovered prior to the great comet hunter recording it. In this case, in 1720, Gottfried Kirch, a German astronomer, predated Messier's find by 62 years.

M5 is a glorious sight and ranks among the three best globulars visible from the southern hemisphere. The 6th magnitude, 12 arc-minute diameter object is about 30 000 light-years distant, corresponding to a linear diameter of 100 light-years.

Small instruments will begin to resolve this richly compressed object. Using 100mm and larger telescopes, many of the resolved stars begin to form patterns, the most notable being 10 or more prominent radiating streams that curve to the extremities. Like Omega Centauri, M5 displays a slight elliptical shape.

Within the same field, just 20 arc minutes to the south, is the double star 5 Serpentis. This uneven pair shows a pretty colour contrast; the yellow 5.2 magnitude primary is separated by 11 arc seconds from its 10th magnitude companion.

Object 70	NGC6121, M4, globular cluster
Constellation	Scorpius
Position	RA 16h20m36s, Dec. −26°24′
Chart number	17

At 7 500 light-years distant, M4 is one of the nearest globulars to our solar system. With a total magnitude of 6.5, this 14 arc-minute diameter star city, populated by at least 20 000 suns, is just visible to the unaided eye under good skies.

A little under 1.5 degrees west of Antares, M4 can easily be located in a finder or binoculars. Telescopes 80 to 100mm of aperture, with medium magnification, will resolve it into many stars, condensing towards the central core. An interesting feature of M4 is the band of 10 fairly bright stars that cross its centre.

Object 71	Alpha Scorpii, Antares, double star
Constellation	Scorpius
Position	RA 16h26m18s, Dec. −26°19′
Chart number	17

Antares, or 'Rival to Mars' was so named by the ancient Greeks owing to its similarity to the red planet.

This red supergiant, the brightest star in

111 OBJECT 60: This almost edge-on spiral galaxy in Centaurus can be seen well in this 20 minute 125mm Schmidt camera photograph. PHOTO: KEN WALLACE

Scorpius, is a difficult but challenging double. Antares, at magnitude 1.2 is separated from its 5.4 magnitude companion by just 3 seconds of arc. The companion sun is drowned by the diffused glare of the primary, not unlike the case with Rigel in Orion.

A good 80mm refractor, steady seeing and high magnification are prerequisites, and will separate the system into a brilliant orange/red star with a very close pretty green companion. The green coloration is a contrast effect, in fact the spectral type of the companion indicates a white star.

Object 72	NGC6405, M6, the Butterfly Cluster
73	BM Scorpii, variable star
Constellation	Scorpius
Position	RA 17h36m48s, Dec −32°11′
Chart number	17

M6 and its close neighbour M7, are undoubtedly two of the finest open star clusters available for the small instrument user.

Known as the Butterfly Cluster, M6 covers an area approximately 25 arc minutes in diameter. The graceful loops and arcs of the brighter members are transformed, with just a little imagination, into the shape of a delightful butterfly with open wings.

Over 80 stars brighter than 11th magnitude are recognised as true cluster members; these are predominantly blue or white. A conspicuous orange semi-regular variable star, designated as BM Scorpii, lies on the extremities of one wing. At its 6th magnitude maximum it is the brightest star in the cluster, fading to the 7th brightest at its 8th magnitude minimum. BM Scorpii has an 850 day period.

Visible to the unaided eye as a small nebulous haze, the cluster is effortlessly resolved in small instruments at low magnification, superimposed against an excellent stellar background of fainter suns. M6 is about 1500 light-years from us.

Object 74	NGC6397, Dunlop 366, globular cluster
Constellation	Ara
Position	RA 17h36m48s, Dec. −53°39′
Chart number	18

NGC6397 is considered to be one of the nearest globulars with a distance of about 6000 light years. At magnitude 7 it makes a fine object for small telescopes, its loose structure allowing easy resolution. Covering an area of 20 arc minutes, the brighter stars seem to form arcs. The globular is very similar in appearance to NGC6121 in Scorpio.

Object 75	NGC6441, Dunlop 557, globular cluster
Constellation	Scorpius
Position	RA 17h46m48s, Dec. −37°02′
Chart number	17

Set in an excellent field, this neat, compact globular makes a fine sight in any telescope, with the brilliant yellow 3.5 magnitude star G Scorpii situated just 4.5 arc minutes away.

The tiny, 3 arc-minute diameter misty globular has an integrated magnitude of 8. With individual stars at magnitude 17, it is beyond resolution. At a distance of 25 000 light-years, NGC6441 is heavily obscured by the interstellar medium. The view through a telescope is reminiscent of NGC5286 in Centaurus.

Object 76	NGC6475, M7, open star cluster
Constellation	Scorpius
Position	RA 17h50m42s, Dec. −34°48′
Chart number	17

About 3.5 degrees from M6 is the large, coarse open cluster M7, set against an extremely rich region of our Milky Way galaxy.

With an estimated distance of 800 light years, M7 sprawls over an area almost 1 degree across. The stars appear mostly blue or white, with the 20 brightest ranging in magnitude from 5.5 to 9.

An integrated magnitude of 5 means that the cluster is easily detected with the unaided eye. The binocular view is superb, the bright members forming a trapezoid. A wide field telescope, or any small instrument at low magnification reveals a beautiful group of 60 or more stars in this, the southernmost object in Messier's catalogue.

By a line of sight coincidence, a small 11th magnitude globular lies within M7's boundaries; this requires excellent skies and a large amateur instrument to detect it.

Overall M7 bears a striking resemblance to its fainter and smaller neighbour M6, although lacking the delicate, intricate character of that cluster.

112 OBJECT 61: NGC5024 is a fine globular cluster in Coma Berenices. This 25 minute exposure with a 250mm telescope shows it well. PHOTO: JOE CAUCHI

Object 77	Munich 15041, Barnard's Star
Constellation	Ophiuchus
Position	RA 17h55m24s, Dec. +04°24′
Chart number	19

Known as Barnard's 'Runaway Star', this red dwarf has, at 5.9 light-years away, the distinction of being our solar system's closest stellar neighbour apart from the Alpha Centauri triple system.

In 1916, from measurements of photographic plates, E.E. Barnard discovered a star with an unusually high angular speed. It displays the greatest proper motion of any star, 10.3 arc seconds per year. Thankfully most stars are too distant to have large proper motions; imagine the devastating effect on star atlases after one or two centuries!

Astronomers have detected slight deviations in the path of Barnard's star, amounting only to a few hundredths of an arc second; these have led to much speculation about invisible planets.

Calculations have established the possibility of two giant planets: an inner planet with a little more mass than Jupiter, orbiting with a period of 11–12 years, at a distance of 3 astronomical units. The outermost, with about half of Jupiter's mass, would take 20–25 years to complete its orbit from about 5 astronomical units from its sun.

There are several other nearby stars that are likely candidates for planetary systems, but as the distance increases and proper motions grow smaller, their detection becomes difficult and uncertain. On a historical note, it may interest readers to know that it was Edmund Halley who is credited with the discovery of proper motions amongst the so called 'fixed stars' in 1718.

Barnard's star is easily located ½° east of 66 Ophiuchi and can be seen as a 9.5 magnitude red star.

Object 78	NGC6514, M20, the Trifid nebula
79	NGC6531, M21, open star cluster
80	HN 40, multiple star
Constellation	Sagittarius
Position	RA 17h58m54s, Dec. −22°30′
Chart number	17

At half the size and just on one and a half degrees to the north of M8, lies the famous Trifid nebula, a vast complex region of hydrogen gas, divided by three radial lanes which inspired the name. Like the Trapezium in the Great nebula in Orion, M20 has a multiple star system, HN40, embedded within.

Colour photographs reveal a beautiful sight: the Trifid displays the purple red of emission nebula, excited by the primary of HN40. Immediately to the north lies a blue reflection nebula, illuminated by the relatively cool star HD164514. The Trifid is probably closer to us than the Lagoon nebula (M8).

HN40 has six components: the main pair, visible in small instruments, magnitude 7 and 8.8, are separated by 10.5 seconds of arc. A third star at 10.5 magnitude and 5.4 arc seconds from the primary may require about 80mm of aperture. The other three stars are only visible in 200–300mm instruments.

The dark lanes can be seen under good dark skies with a 60mm refractor, while 100mm telescopes and over uncover abundant detail.

In a low power field, the open star cluster M21 can be observed just over half a degree to the northeast of M20. It is a compact group with about 40 members brighter than 12th magnitude, six of them being 8th magnitude. Due to the rich stellar background, it is difficult to define the exact limits of M21. A yellow star in the centre is contrasted by the predominantly white cluster members.

While there is a possible connection between the Trifid and Lagoon, at a distance of 4000 to 5000 light-years, M21 lies much closer to us at 2200 light-years.

Object 81	NGC6523, M8, the Lagoon nebula
82	NGC6530, open star cluster
Constellation	Sagittarius
Position	RA 18h01m36s, Dec. −24°20′
Chart number	17

The Lagoon nebula is a complex HII region, infiltrated by irregular lanes and prominent dark globules. The very young star cluster NGC6530 lies embedded in the half-degree diameter nebulosity, adding appreciably to its visual appearance.

The bright young members of NGC6530 are chiefly responsible for illuminating M8. Near the brightest region of nebulosity is the 9th magnitude star Herschel 36, named after its discoverer, Sir John Herschel. H36 is one of the youngest stars known; probably less than 10 000 years have elapsed since its birth from the nebula. Alongside and associated with H36 is the tiny, very bright, 30 arc-second diameter emission nebula known as the Hourglass.

Crossing the centre of M8 is a dark lane of obscuring material, which gave rise to the name Lagoon. Many globules are silhouetted against the bright nebulous background. These dark, compact areas are thought to be compressed concentrations of dust that may one day collapse into new stars.

With an integrated magnitude of 5, M8 is visible to the unaided eye under good dark skies. Set in a glorious portion of the Milky Way, the binocular view is splendid with M20 located about one and a half degrees to the north. Spend a few moments with binoculars and sweep south-west of M8 into Scorpius, the view of the starfield is rewarding.

Any size telescope at low magnification will show many fine features, including the dark lagoon rift, in this celestial showpiece. The distance is uncertain, 4000 to 5000 light-years seems favoured.

113 OBJECT 62: This fine shot of NGC5128 shows not only the galaxy, but also a supernova which was discovered in the dust lane on 4 May 1986 by Australian amateur astronomer Robert Evans. The supernova can be seen as the bright star at the right hand end of the dust lane. A 25 minute exposure with 250mm telescope produced this fine shot. PHOTO: JOE CAUCHI

Object 83 70 Ophiuchi, double star
Constellation Ophiuchus
Position RA 18h02m54s, Dec. +02°31′
Chart number 19

This double is a real gem, comprised of two dwarf stars situated just fifteen light-years from our solar system. The distance between the pair is about one third of a light-year and their period is 87.8 years. Seen from Earth their separation varies from about 1.7 to a maximum of 6.7 seconds of arc; their magnitudes are 4.3 and 6.

As is the case with many doubles with colour contrast, observers tend to give varied estimates of their hue. 70 Ophiuchi is no exception with yellow and red, topaz and violet having been quoted. I see them as yellow and orange.

An 80mm refractor should split the pair, but as their separation is at minimum during the 1980s, high magnification and good seeing is a prerequisite.

114 OBJECT 63: NGC5139 or ω Centauri is the finest globular cluster in the sky. Lying at a distance of 16 500 light-years it appears over 30′ of arc wide. This 10 minute exposure with a 250mm telescope shows it well. PHOTO: JOE CAUCHI

Object 84 NGC6611, M16, open star cluster
85 IC4703, the Eagle nebula
Constellation Serpens
Position RA 18h16m00s, Dec. −13°48′
Chart number 20

The open star cluster M16 contains about 60 predominantly blue and white stars, the 25 brightest ranging in magnitude from 8 to 11. The cluster resides within the splendid nebula known as the Eagle.

The older members of the cluster are estimated to be young, 2 million years old, and within the depths of the Eagle nebula, stellar formation is undoubtedly still in progress.

Photographically the Eagle nebula is a grandiose sight: faint and bright stars stud the vivid HII region, which covers an area 35 by 28 minutes of arc. Dark globules and intrusions are common and may ultimately condense into stars. The overall impression is that of an Eagle with spread wings.

The view through a small telescope is excellent, with the 20 arc-minute wide cluster immersed in a nebulous haze; a 100mm telescope shows three distinct nebulous regions. The distance is 4500 light-years.

Object 86 NCG6618, M17, the Omega nebula
Constellation Sagittarius
Position RA 18h18m00s, Dec. −16°12′
Chart number 20

Noted by Charles Messier in 1764, this bright, conspicuous nebula with an integrated magnitude of 6, presents an interesting and easy subject for the modest telescope.

The shape of M17 has given rise to names like the Omega, Horseshoe and Swan nebula. The brightest portions in a small instrument measure about 15 by 5 arc minutes, and resemble a figure 2 with an exaggerated baseline. The Omega and Horseshoe figures are only half completed as the 2 does not form an opposing mirror image. The Swan, however, is easily discernible, with its curved neck and bold body, adrift amidst a fine field of stars.

A cluster of about 35 stars is associated with M17, but is inconspicuous in a small telescope due to the star-rich nature of the field. The distance to M17 is a little over 3000 light-years.

115 OBJECT 64: NGC5236 is a fine, face-on spiral galaxy in Hydra. The photograph was taken with a 200mm telescope and 30 minute exposure. PHOTO: KEN WALLACE

Object 87	NGC6633, open star cluster
Constellation	Ophiuchus
Position	RA 18h25m06s, Dec. +06°32′
Chart number	21

An interesting open cluster with about 65 members of 7th magnitude down. The 'A' shape is somewhat reminiscent of NGC4755, the Jewel Box in Crux, but considerably larger with a diameter of 20 minutes of arc.

The cluster is 1600 light-years from our solar system and has a diameter of about 10 light-years.

Object 88	NGC6656, M22, globular cluster
Constellation	Sagittarius
Position	RA 18h33m18s, Dec. −23°57′
Chart number	20

Described by Sir John Herschel as a noble object, M22 is surpassed in brightness only by the globulars Omega Centauri and 47 Tucanae. Only the fact that its light is dimmed by intervening interstellar dust prevents it from being a gem of the southern skies.

With an integrated magnitude of 6, M22 is just visible to the unaided eye. At a distance of 10 000 light-years, it is one of the nearest globulars and contains upwards of 70 000 stars within its 50 light-year diameter.

The 18 arc-minute wide M22 is an ideal subject for the small instrument; its 11th magnitude stars resolve easily out from its compressed centre.

Object 89	NGC6705, M11, the Wild Duck cluster
Constellation	Scutum
Position	RA 18h48m24s, Dec. −06°20′
Chart number	20

The eleventh object in Messier's catalogue, the Wild Duck cluster, is one of the most populous of the open star clusters known, with over 600 stars ranging from 9th down to 16th magnitude, all compressed into a 12 arc-minute diameter ball. At a distance of 5500 light-years M11's linear diameter is about 20 light-years. The night sky from a planet circling around an M11 sun would be a dazzling blaze of brilliant points of light.

With an integrated magnitude of 6, M11 is easily detected with the unaided eye as a concentrated patch of light, in the wonderfully rich region of the Milky Way known as the Scutum Star Cloud. Through binoculars and small telescopes it resembles a globular. A 60 to 100mm telescope at medium magnification will resolve many stars, and reveal the distinctive wedge shape that prompted Admiral Smyth to compare it to wild ducks in flight.

Object 90	Theta Serpentis, Struve 2417, double star
Constellation	Serpens
Position	RA 18h53m48s, Dec. +04°08′
Chart number	21

A star-sprinkled field makes an attractive setting for this celestial gem. The two white suns, 4.5 and 5.4 magnitude, separated by 23 seconds of arc, are easily split in the smallest of instruments. Prominent amidst the field is an orange star, situated 7 minutes of arc from the pair.

Object 91	NGC6752, Dunlop 295, globular cluster
Constellation	Pavo
Position	RA 19h06m24s, Dec. −60°04′
Chart number	22

An excellent globular cluster that is often neglected by amateurs. With its southerly declination, it is not surprising that is omitted, or only glossed over by northern hemisphere astronomy books. Many southern amateurs rank this globular among the top 3, surpassed only by 47 Tucanae and Omega Centauri.

The angular size of NGC6752 is quite large, 15 minutes of arc. The linear diameter is 100 light-years, and it takes 20 000 years for its light to reach us.

Although located in a sparse field in Pavo, it is easily located in finder scopes. A small telescope with moderate magnification reveals a diffuse cloudlike patch, studded with stars. Emanating from the condensed centre are impressive strings of stars of varying magnitudes, curving out in arc-like arms. One Australian amateur, Mark Fisher, appropriately nicknamed NGC6752 the 'Starfish Cluster'.

James Dunlop discovered this globular in the 1820's, at Parramatta N.S.W. His interpretation was that of two objects, one close and resolvable, the other very distant, in the same line of sight and unresolvable. Indeed the bright stars against the nebulous background can give this impression.

Object 92 Beta Sagitarii, Dunlop 226, double star
Constellation Sagittarius
Position RA 19h19m00s, Dec. −44°33′
Chart number 23

A dainty pair set in an excellent field. The 4.2 magnitude blue–white primary is separated from its 7th magnitude white companion by 28 arc seconds, making it an easy target for small apertures.

Object 93 Epsilon Sagittae, double star
Constellation Sagitta
Position RA 19h35m00s, Dec. +16°21′
Chart number 21

A good, wide double suitable for the smallest of apertures. The suns, 5.7 and 7.7 magnitude, are separated by almost 90 arc seconds. In agreement with their spectral class, G8 and B8, they appear as a pretty yellow and blue. Forming a triangle with the main pair is a faint 12th magnitude companion.

Object 94 HN84, Hh630, double star
Constellation Sagitta
Position RA 19h37m12s, Dec. +16°27′
Chart number 21

Another wide double with fine colour contrast, almost in the same low power field as Epsilon Sagittae. The orange–red and blue stars, of magnitude 6.5 and 8.6, are separated by 28 arc seconds.

116 OBJECT 68: NGC5904, or M5, is a fine globular cluster in Serpens. This 25 minute exposure with a 250mm telescope shows the star 5 Serpentis just above it. PHOTO: JOE CAUCHI

Object 95 Dunlop 227, double star
Constellation Telescopium
Position RA 19h48m42s, Dec. −55°06′
Chart number 23

An excellent pair with colour contrast. The stars are similarly matched at 6.2 and 6.8 magnitude; separation is an easy 23 arc seconds. The colour contrast is good with the brighter star golden yellow, and the companion sun white.

Object 96 NGC6838, M71, globular cluster
Constellation Sagitta
Position RA 19h51m30s, Dec. +18°39′
Chart number 21

The exact nature of M71 has been the subject of some controversy. It is not clear if it is a very compressed open cluster as indicated by its metal-abundant stars, or a globular as other observations suggest.

In a small instrument, this 6 arc-minute diameter object, with an integrated magnitude of 8, appears as a nebulous oval, with one side brighter than the other. Telescopes in the 100 to

117 OBJECT 72: NGC6405, or the Butterfly cluster, is one of the prettiest open star clusters in the sky. This 20 minute exposure with a 20mm telescope shows the shape of the cluster well. PHOTO: KEN WALLACE

150mm aperture range will begin to resolve some stars, but noticeable is the lack of a compressed centre, common to most globulars. The distance is also uncertain, between 8000 and 18000 light-years.

Object 97 NGC7009, the Saturn nebula
Constellation Aquarius
Position RA 21h01m24s, Dec. −11°34′
Chart number 24

This 8th magnitude blue–green planetary appears stellar in small telescopes. High magnification will show the 25 second of arc disc to be slightly oval. Dark skies and at least a 300mm telescope are required to glimpse the ansae which project from each side, resembling the planet Saturn with its rings edge-on. This distance is estimated at about 4000 light-years.

Object 98 Theta Indi, Herschel 5258, double star
Constellation Indus
Position RA 21h16m18s, Dec. −53°40′
Chart number 23

This is a beautiful double in a fine field. The 4.6 magnitude yellowish primary is set 6 seconds of arc from its 7.2 magnitude red companion sun. Small apertures with medium magnification will show this pair well.

Object 99 NGC7089, M2, globular cluster
Constellation Aquarius
Position RA 21h30m54s, Dec. −01°04′
Chart number 24

This beautiful globular, number 2 in Messier's catalogue, is found in a rather sparse field. At magnitude 6 it is an easy object in binoculars, but will require a telescope of a least 100mm diameter to begin to resolve its very condensed 8 arc-minute diameter centre, and to detect its outlying rays of stars. The light from its 100000 suns, compressed over 150 light-years, has taken 50000 years to reach us.

Object 100 NGC7099, M30, globular cluster
Constellation Capricornus
Position RA 21h37m30s, Dec. −23°25′
Chart number 24

Number 30 in the catalogue of Messier, this 8th magnitude, 6 arc-minute diameter globular lies in a good field, with the 5.5 magnitude star 41 Capricorni just 25 arc minutes away. M30's distance is estimated at 40000 light-years, with a diameter of 75 light-years.

Small instruments show a distinct fuzzy core, surrounded by a haze of unresolved stars. With 100mm of aperture a number of stars become visible in the outer regions.

Object 101 NGC7293, the Helix nebula
Constellation Aquarius
Position RA 22h27m00s, Dec. −21°06′
Chart number 1

The Helix, sometimes called the Helical nebula, is the largest and possibly the nearest of the planetaries, but like most of this class of object its distance is not accurately known. Its integrated magnitude is 6, but due to its large half degree size (the same area that the moon covers!), its surface brightness is very low, making a difficult object for the novice. It is best viewed with binoculars under good dark skies, and appears as a large circular hazy patch. A small telescope should not have any trouble; just use the lowest magnification possible, and remember it covers a large area.

Object 102 Beta Piscis Austrini, Dunlop 240, double star
Constellation Piscis Austrinus
Position RA 22h28m42s, Dec. −32°36′
Chart number 1

A splendid double with a contrasting magnitude difference, Beta Piscis Austrini has magnitudes 4.4 and 7.8, and separation of a wide 30 seconds of arc. The yellow and white pair are easily resolvable in the smallest of instruments.

Object 103 h5356, multiple star
Constellation Piscis Austrinus
Position RA 22h37m00s, Dec. −28°36′
Chart number 1

h5356 is a good multiple star system, described by Sir John Herschel as elegant. The 6.3 magnitude yellow primary is set on its own, 87 arc seconds away from a fainter pair at magnitude 7.3 and 8.1, separated by a close 3 arc seconds. High magnification may be required to separate the faint pair.

118 OBJECT 75: NGC6441 is a small globular cluster in Scorpius. The 3.5 magnitude star G Scorpii can be seen just 4.5′ away from the globular. The six points on the star are caused by diffraction effects around the secondary mirror holder in the 200mm telescope used for this 20 minute exposure. PHOTO: KEN WALLACE

119 OBJECT 76: NGC6475 is one of the many open star clusters in Scorpius. This photo is a 27 minute exposure using a 250mm telescope. PHOTO: JOE CAUCHI

120 OBJECTS 78 AND 79: The Trifid nebula in Sagittarius is a fine example of a complex region of gas and dust where stars are forming. At the lower right corner the open cluster M21 can be seen. This 25 minute 250mm shot shows both objects to full advantage. PHOTO: JOE CAUCHI

121 OBJECT 81: NGC6523, the Lagoon nebula, is another example of a gas cloud in Sagittarius. This 22 minute exposure with a 250mm telescope shows much of the detail in the nebula and its dark lane of dust. PHOTO: JOE CAUCHI

122 OBJECT 85: IC4703 is also known as the Eagle nebula. Use your imagination and you can just make out an eagle diving on its prey. The photo is a 25 minute exposure with a 250mm reflector. PHOTO: JOE CAUCHI

123 OBJECT 86: NGC6618 is also known as the Omega nebula because of its resemblance to half of the Greek letter. A 25 minute shot with a 250mm reflector. PHOTO: JOE CAUCHI

124 OBJECT 88: NGC6656, a globular cluster in Sagittarius, is one of the brightest of the globulars. This 18 minute exposure with a 250mm reflector shows the stars in the outlying regions quite well. PHOTO: JOE CAUCHI

125 OBJECT 89: NGC6705, or the Wild Duck cluster, is one of the most populous open clusters known. Over 600 stars are in the cluster, many of them visible in this 18 minute, 250mm photograph. PHOTO: JOE CAUCHI

126 OBJECT 96: NGC6838 is a small globular cluster in Sagitta. An 11 minute exposure with a 250mm reflector. PHOTO: JOE CAUCHI

127 OBJECT 99: NGC7089, or M2, is a fine globular cluster in Aquarius. It is a 6th magnitude object which needs at least 100mm of aperture to see the outlying stars. This 10 minute 250mm photo shows them easily. PHOTO: JOE CAUCHI

128 OBJECT 101: NGC7293, the Helix nebula, is one of the most fascinating of the planetary nebulae. This 25 minute, 250mm photograph shows much of the structure visible in large telescopes. Even in small telescopes the Helix is an amazing object. PHOTO: JOE CAUCHI

Maps and charts

HOW TO USE THE SEASON ASPECT MAPS

The season aspect maps detail the principal constellations visible from the southern hemisphere. When used in conjunction with the 'Hour of right ascension on the meridian' table (9.2), you will be able to identify the constellations for any time throughout the year.

The maps represent the appearance of the sky on the following dates:

Spring	6 October at 9.00pm (22h on the meridian)
Winter	7 July 9.00pm (16h on the meridian)
Autumn	6 April 9.00pm (10h on the meridian)
Summer	5 January 9.00pm (04h on the meridian)

Each map has two parts: the rectangular part shows the stars which rise and set, the circular part shows those stars which are always visible, the circumpolar stars.

As declination is related to latitude on Earth, the exact orientation of the maps against the stars will depend upon your position. If you live on the equator, latitude 0°, stars of 0° declination cross the meridian directly overhead, and the north and south celestial poles will be on the northern and southern horizons, 90° from the celestial equator. If you live in Adelaide, South Australia, with latitude 35°S, stars with declinations of −35° will transit the meridian at the zenith. Your northern limit will be 35° south of the NCP, or 55° north of the celestial equator. The SCP will be 35° above your southern horizon and any star with declination south of −55° will be circumpolar.

As you travel further south the higher the SCP appears in the sky. Once you reach the Earth's south pole the SCP will be directly overhead, 0° declination will be on the horizon and no northern stars will be visible. To use the season aspect maps for dates other than those shown above, look up table 9.2 for the appropriate time and date. Locate the aspect map with this RA. Holding the map overhead with the top towards north, all constellations with increasing RA will be seen west of the meridian and those with decreasing RA to the east.

The circumpolar map need only be rotated so the appropriate RA is at the top of the circle when you are facing south.

129 Symbols used on charts

Stellar magnitudes

1+

1–2

2–3

3–4

4–5

5–6

6–7

Objects

Globular Star Cluster

Galaxy

Planetary Nebula

Diffuse Nebula

Open Star Cluster

Identified with catalogue number, NGC etc.

Double stars identified with Greek Letter, and or discoverer.

h = Sir John Herschel

Δ = James Dunlop

Σ = F. Struve

Milky Way and Magellanic Clouds

Constellation boundaries

Constellations are identified with their standard three letter abbreviation, eg Centaurus = Cen

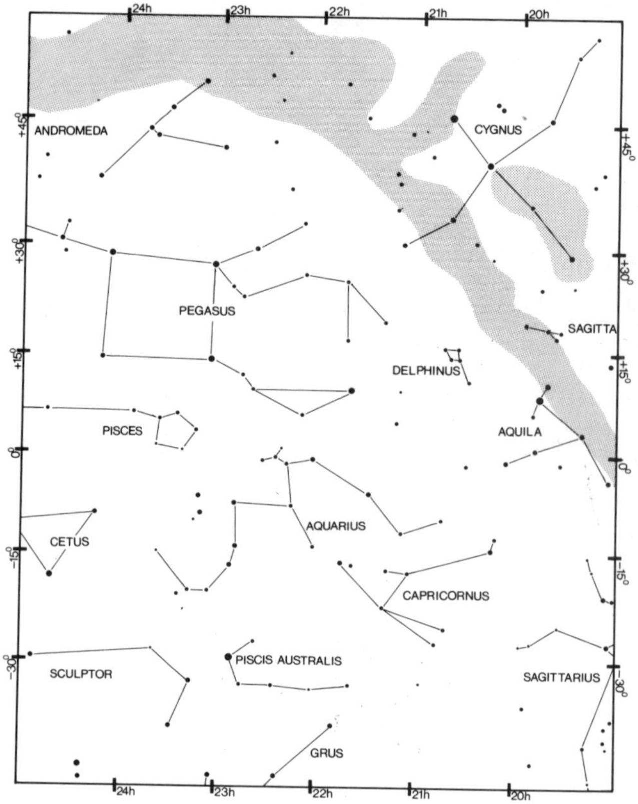

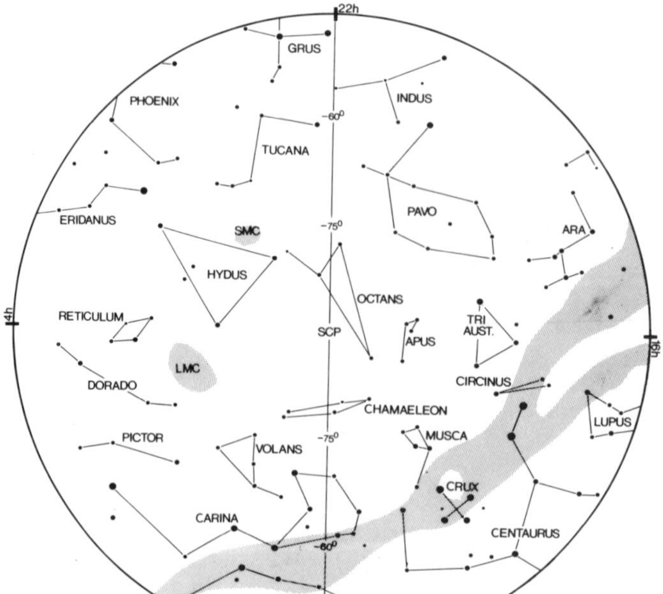

130 Spring aspect maps

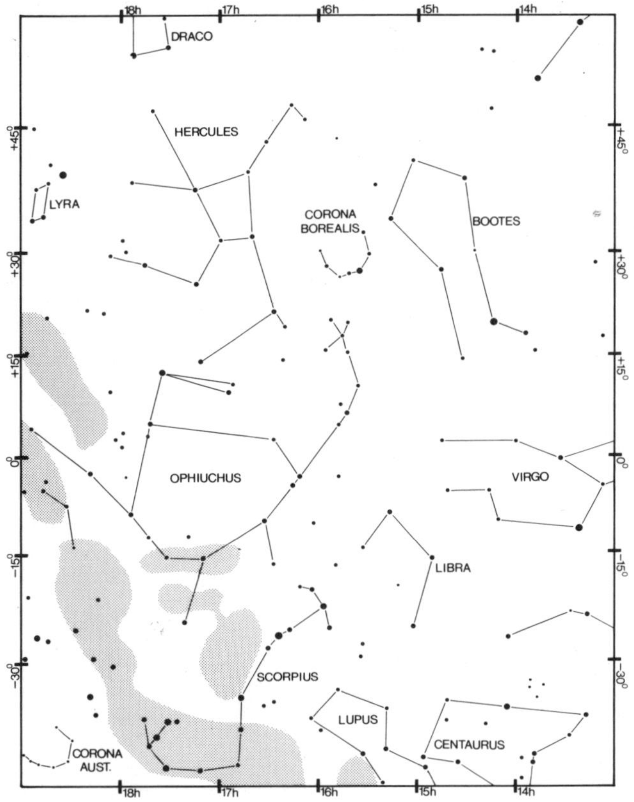

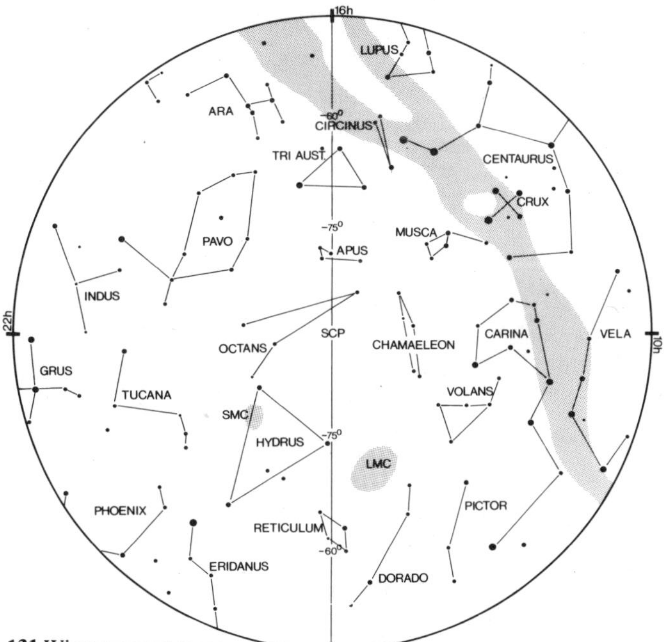

131 Winter aspect maps

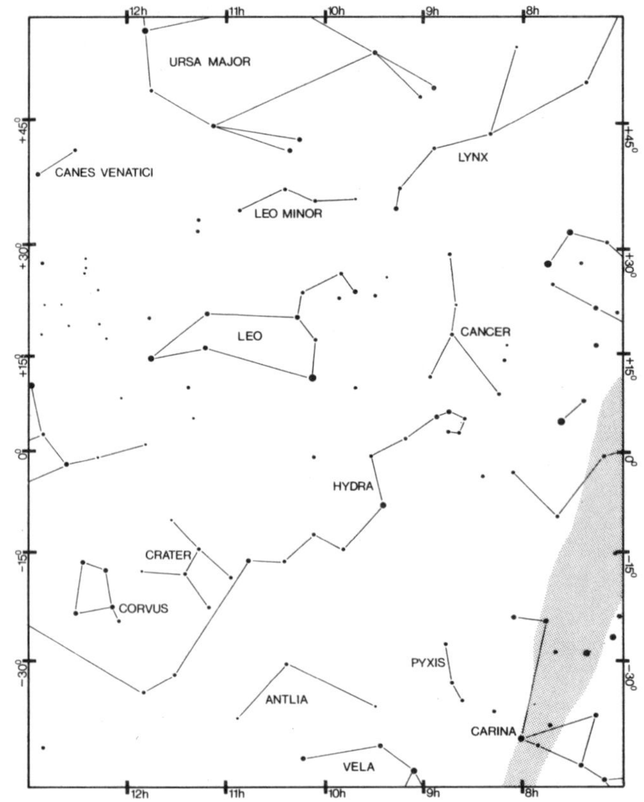

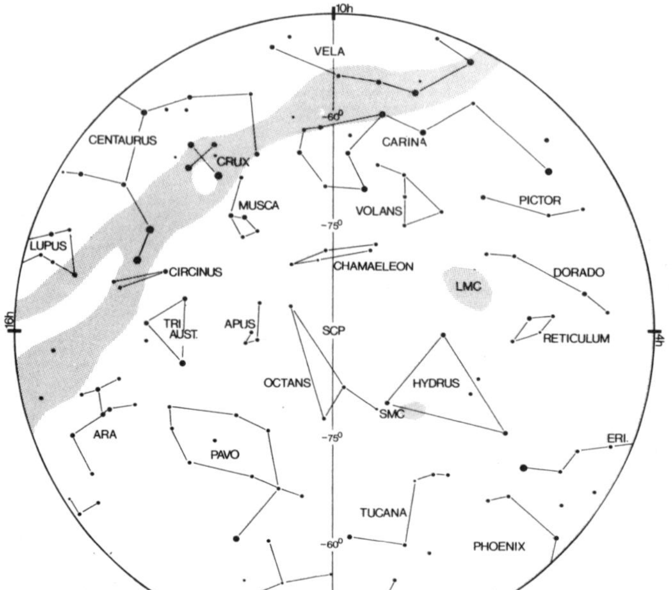

132 Autumn aspect maps

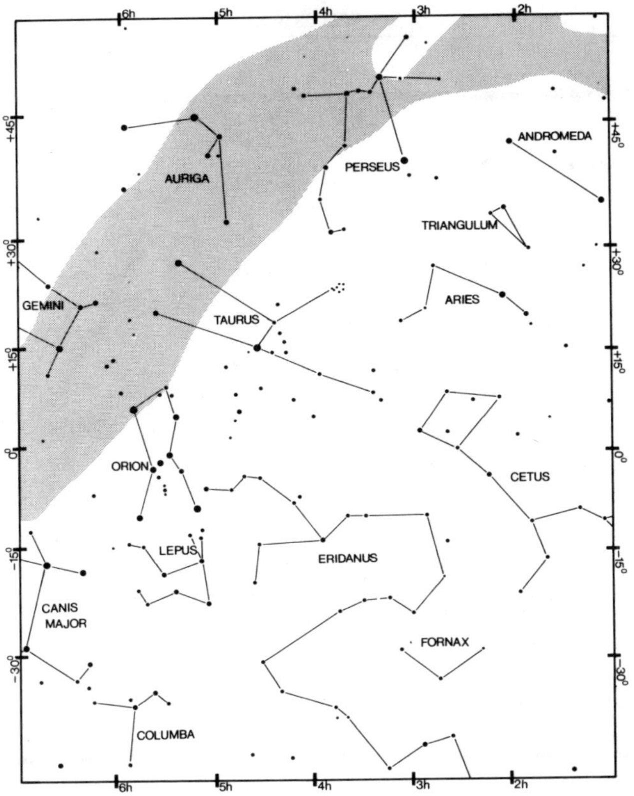

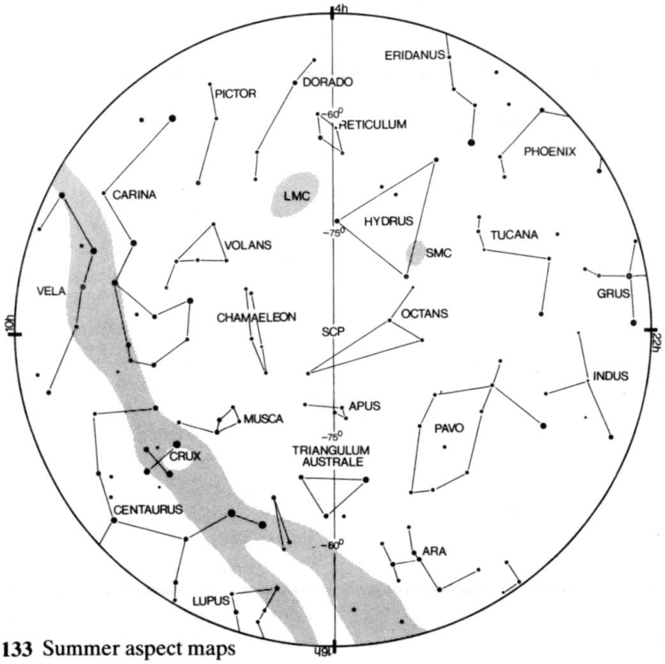

133 Summer aspect maps

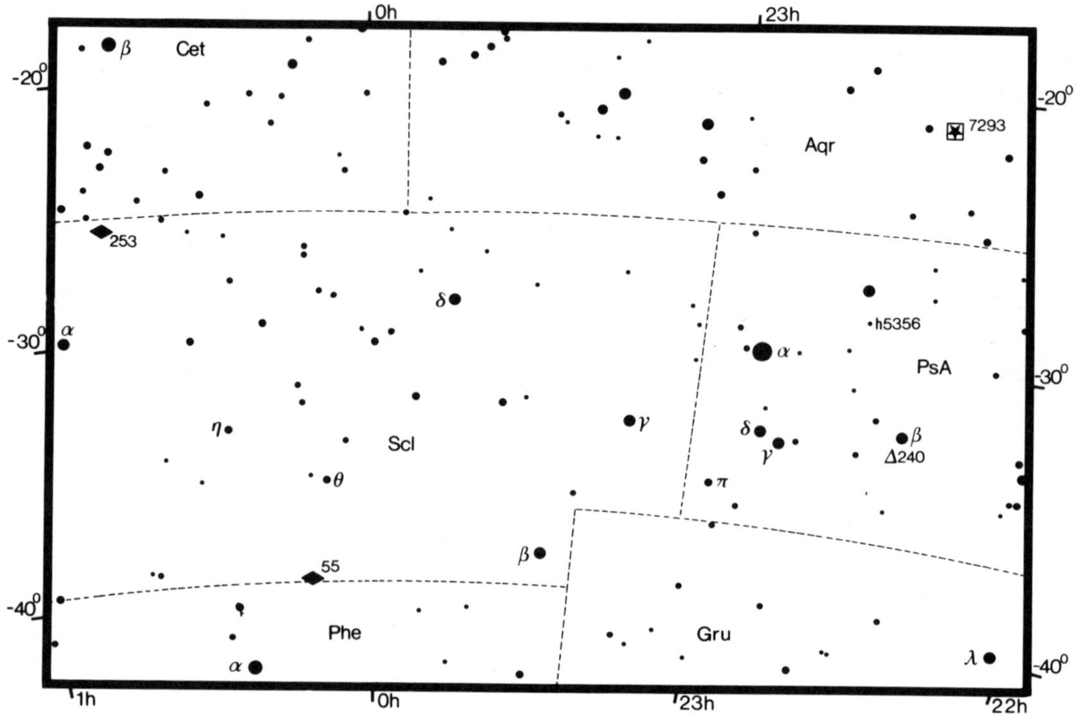

134 Object finder chart 1

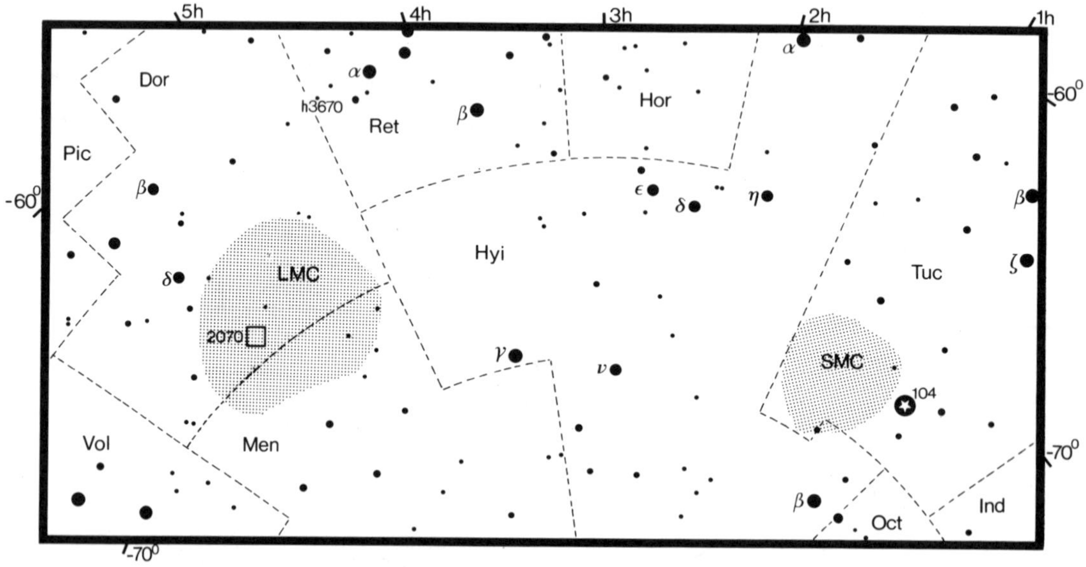

135 Object finder chart 2

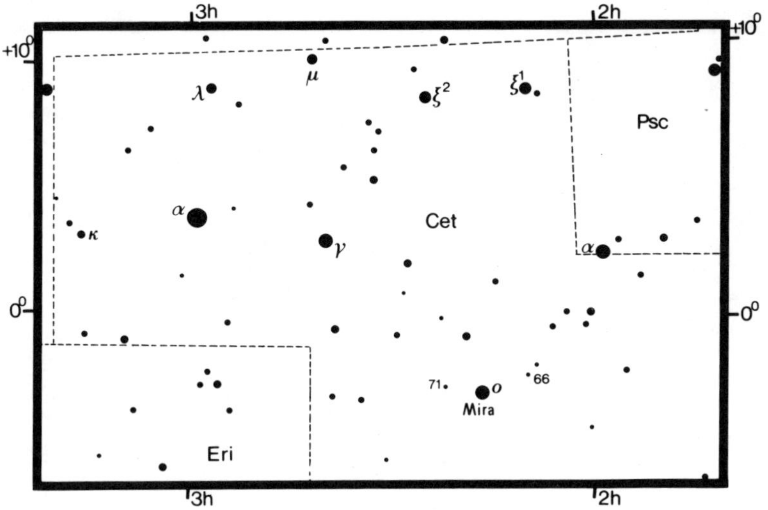

136 Object finder chart 3

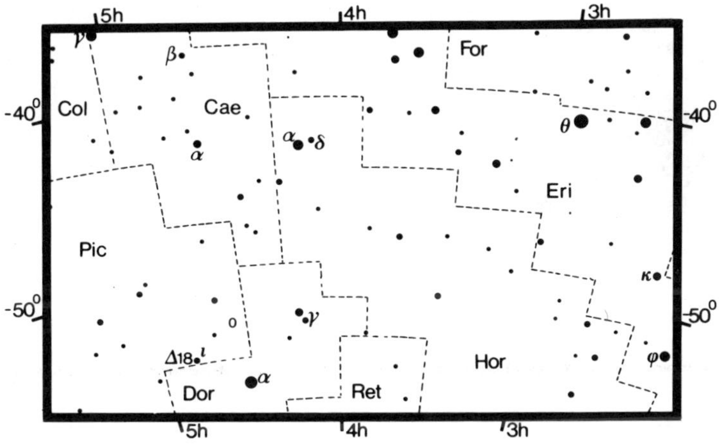

137 Object finder chart 4

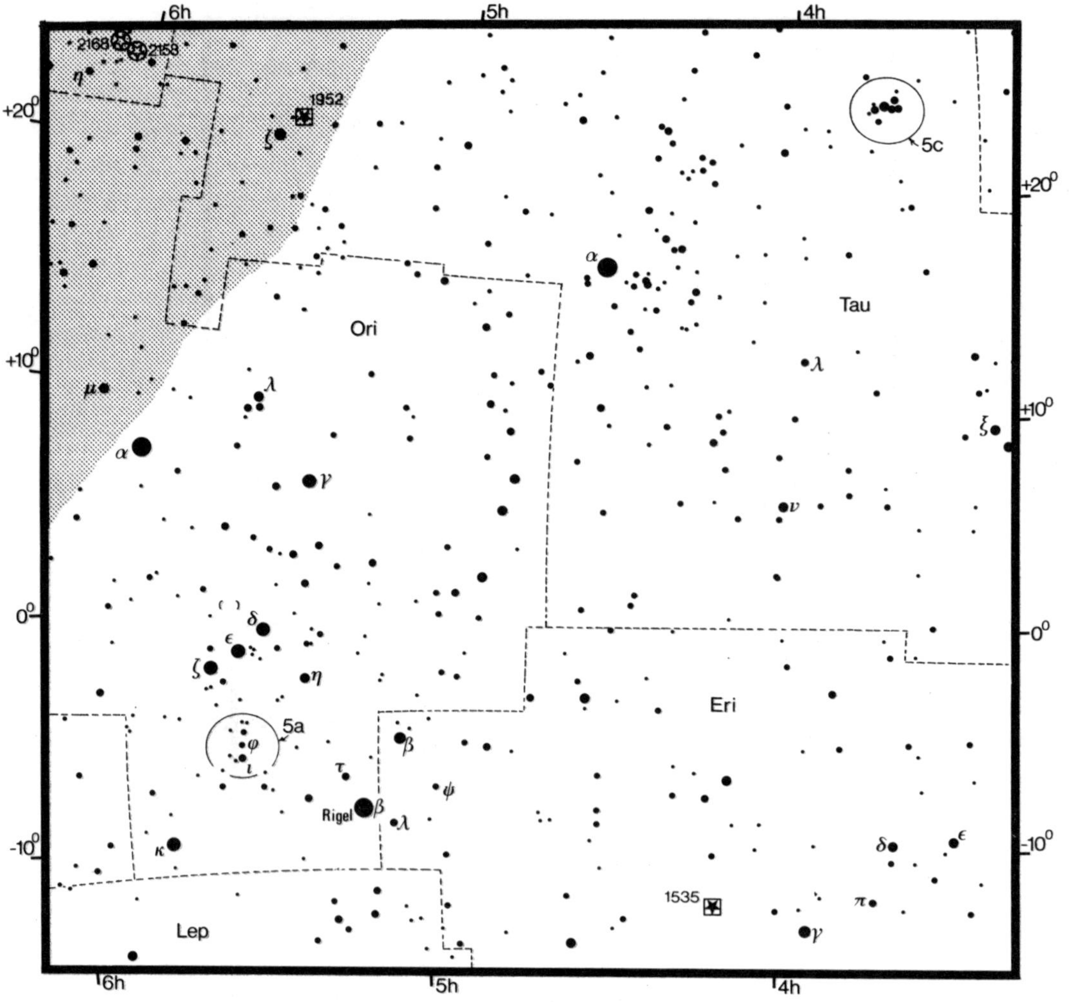

138 Object finder chart 5

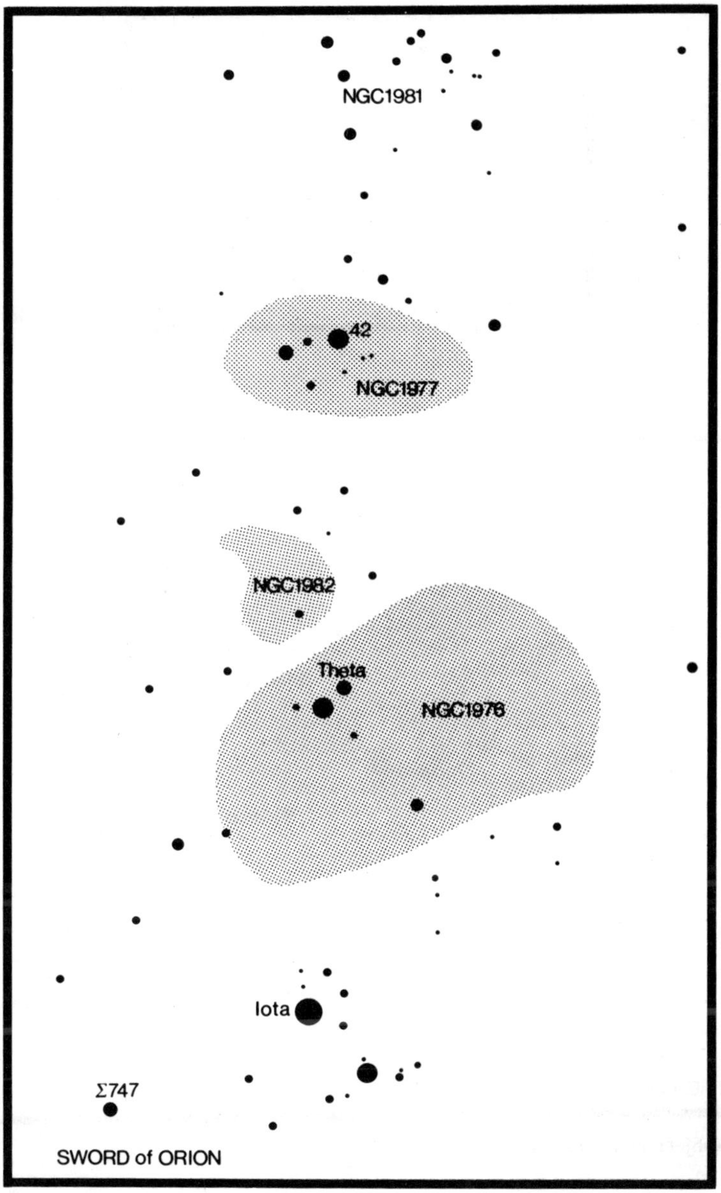

139 Object finder chart 5a

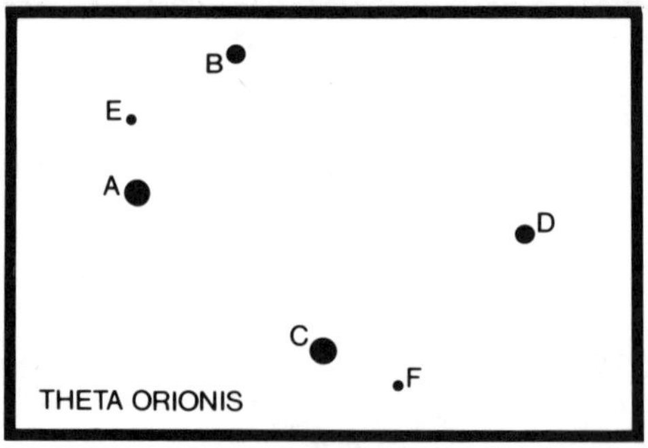

140 Object finder chart 5b

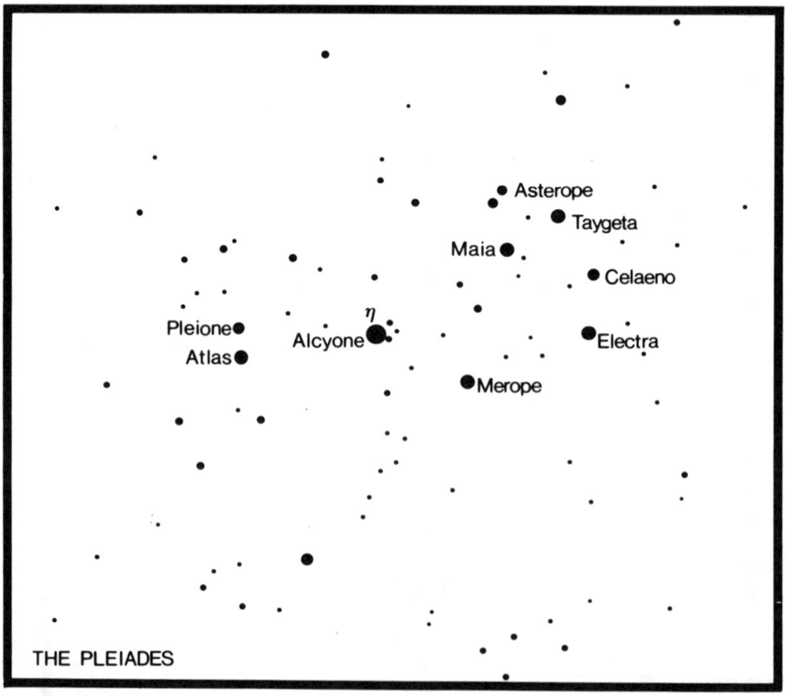

141 Object finder chart 5c

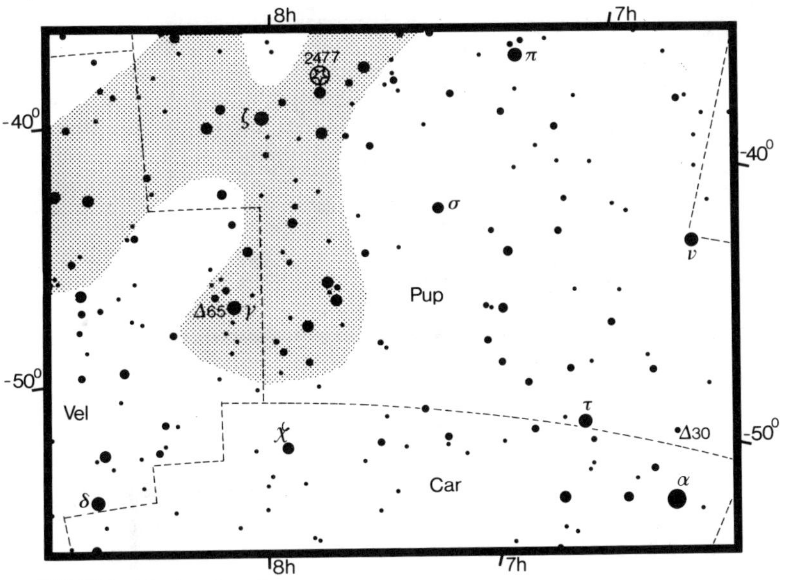

142 Object finder chart 6

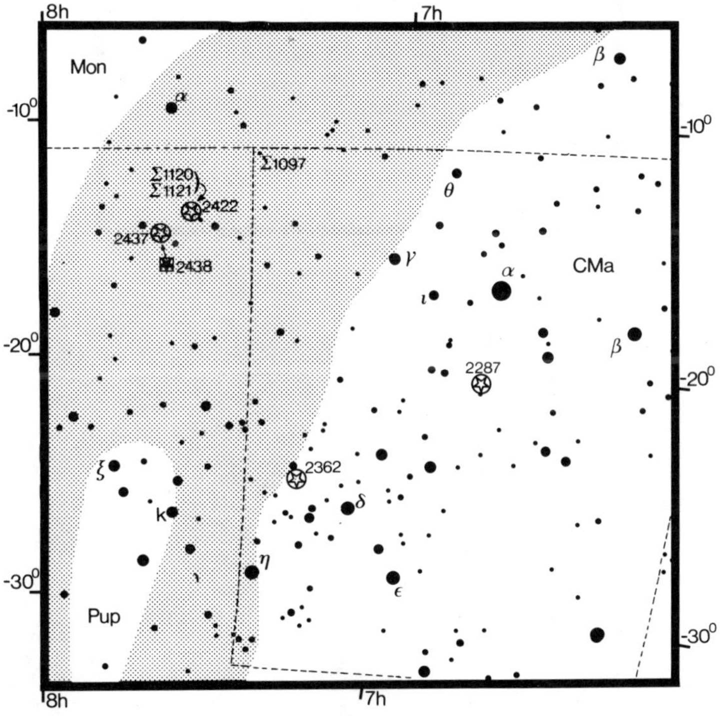

143 Object finder chart 7

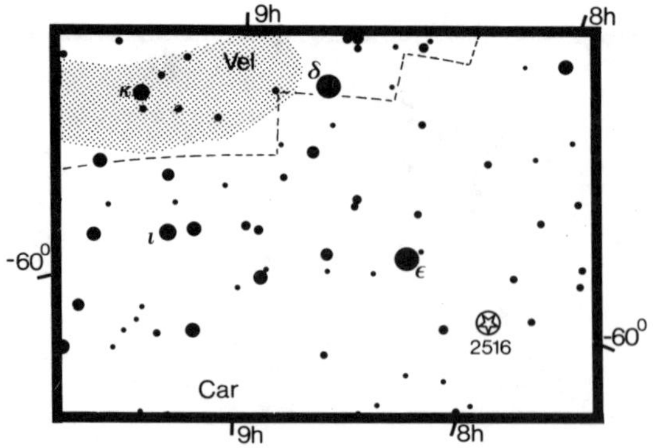

144 Object finder chart 8

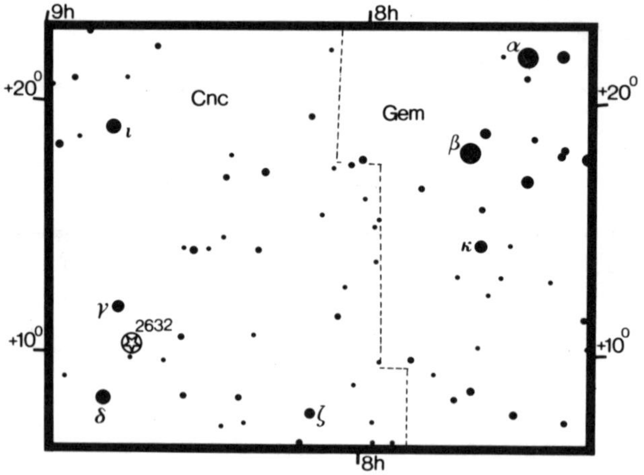

145 Object finder chart 9

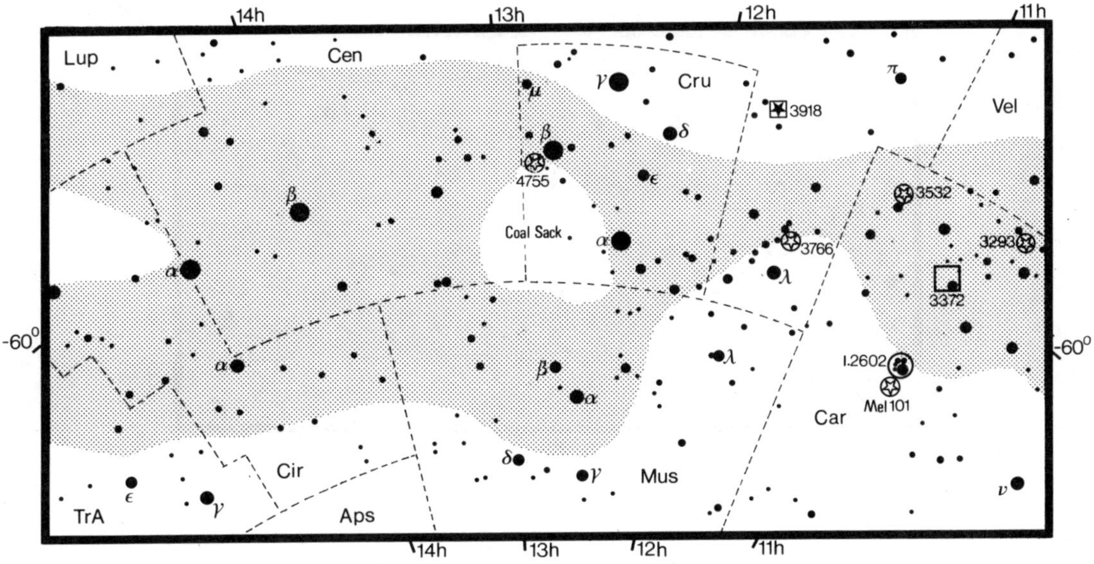

146 Object finder chart 10

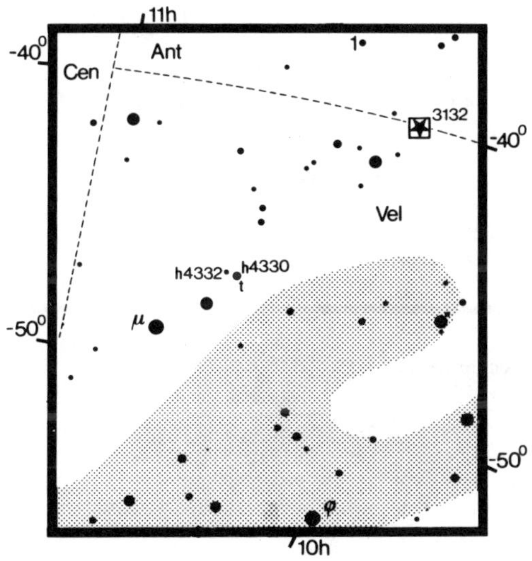

147 Object finder chart 11

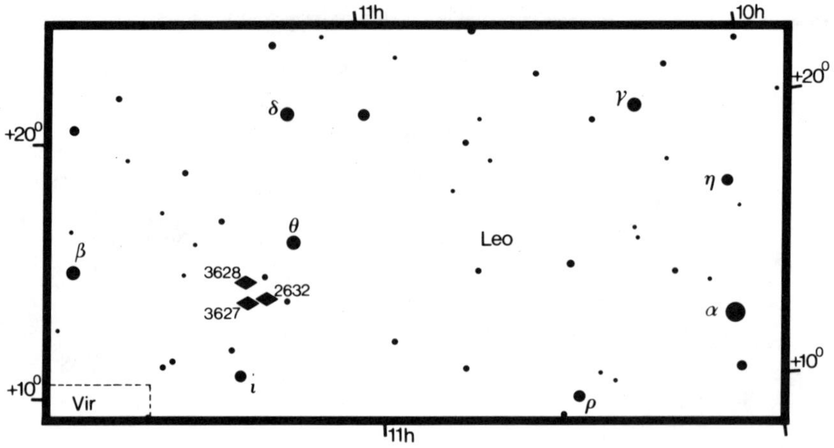

148 Object finder chart 12

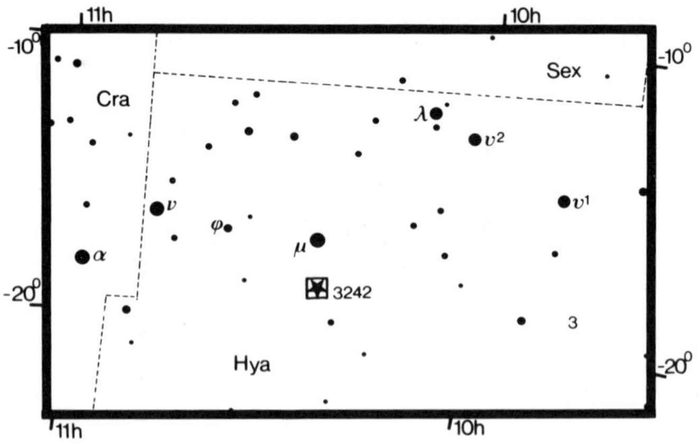

149 Object finder chart 13

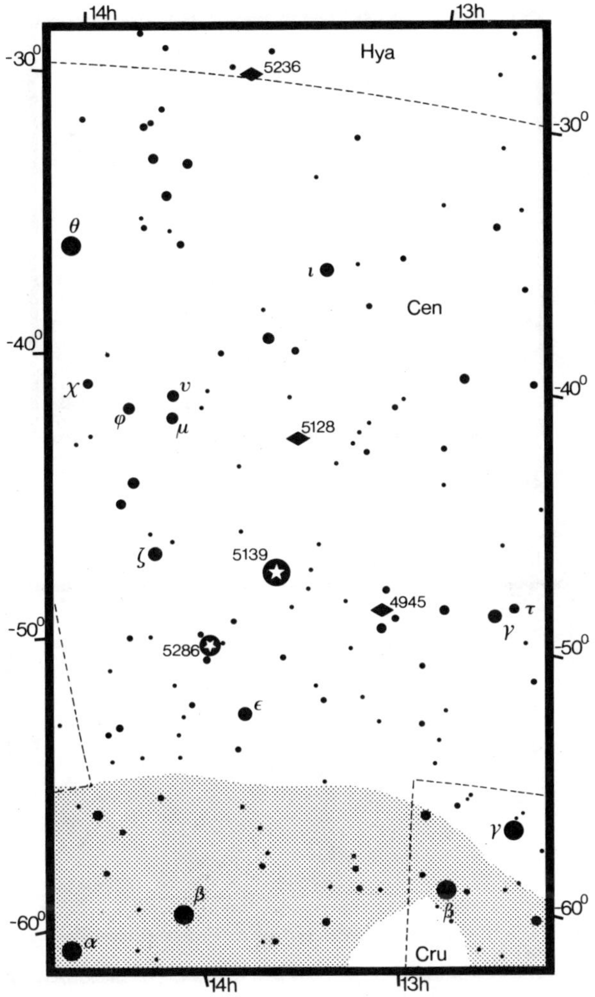

150 Object finder chart 14

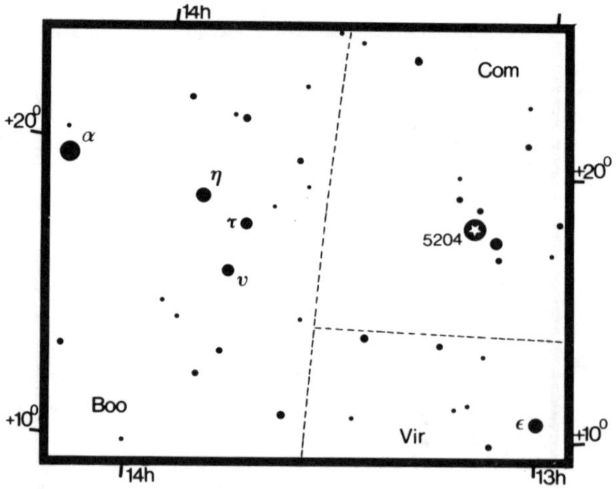

151 Object finder chart 15

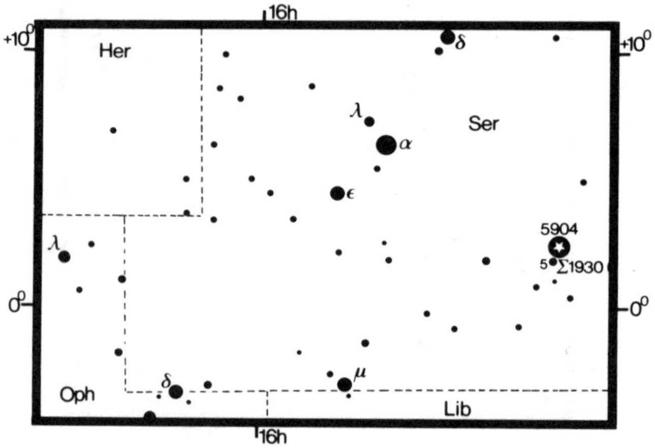

152 Object finder chart 16

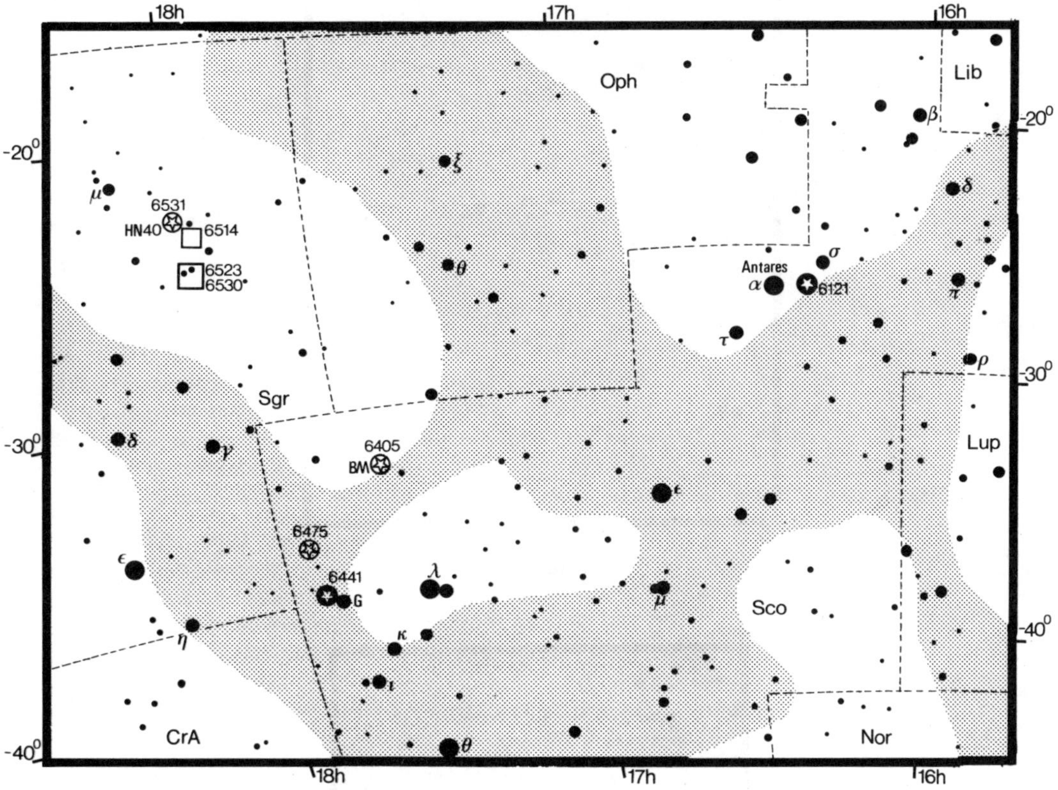

153 Object finder chart 17

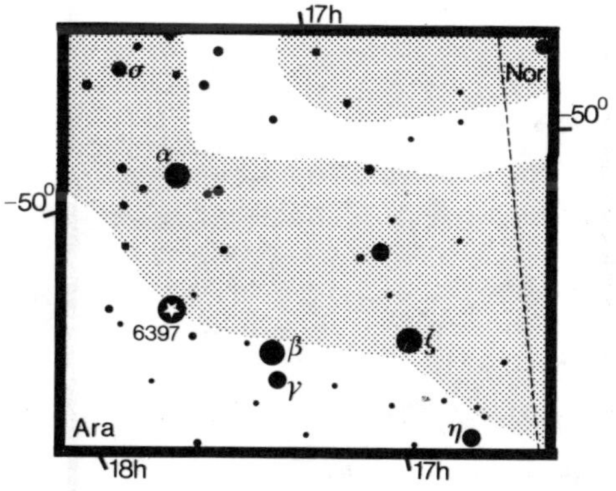

154 Object finder chart 18

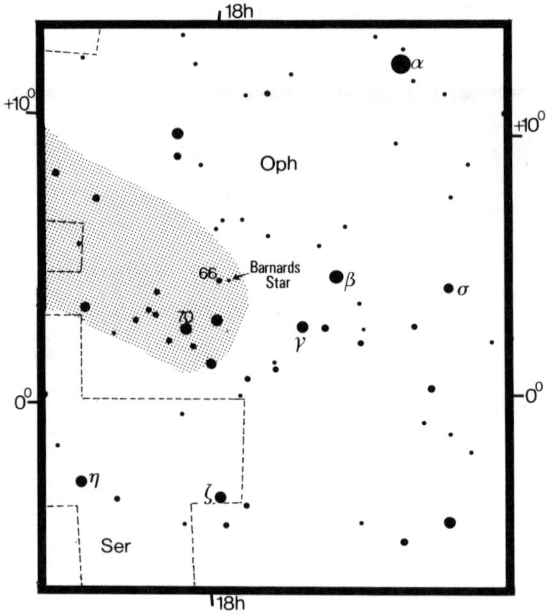

155 Object finder chart 19

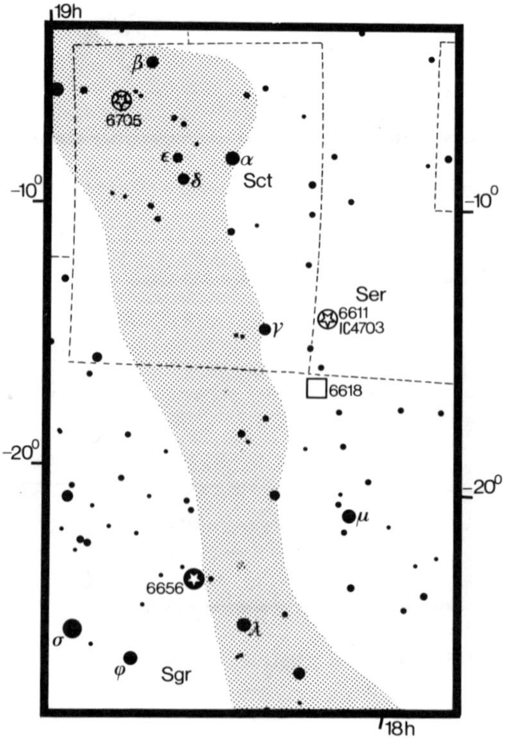

156 Object finder chart 20

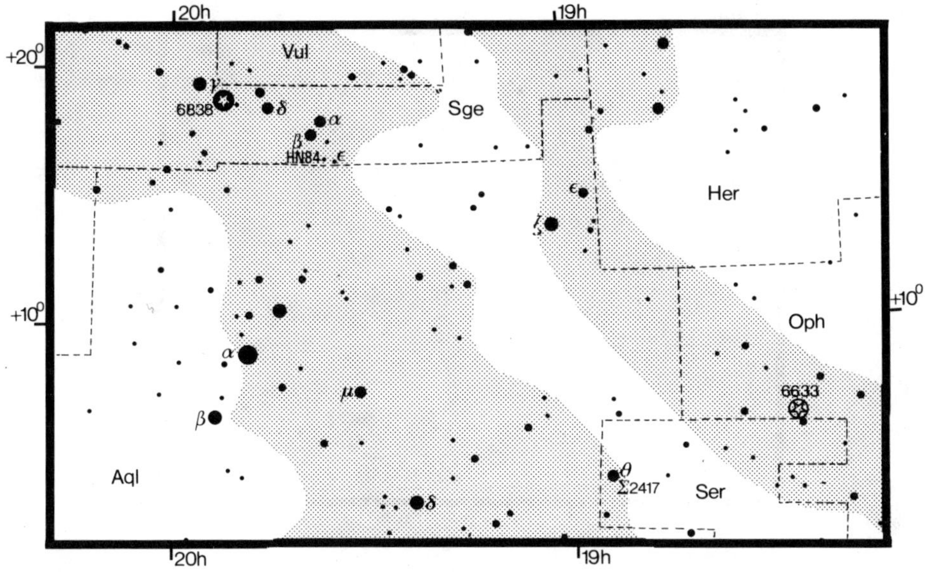

157 Object finder chart 21

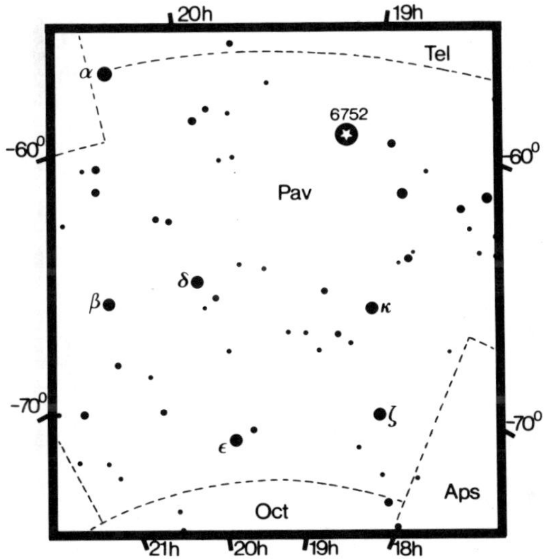

158 Object finder chart 22

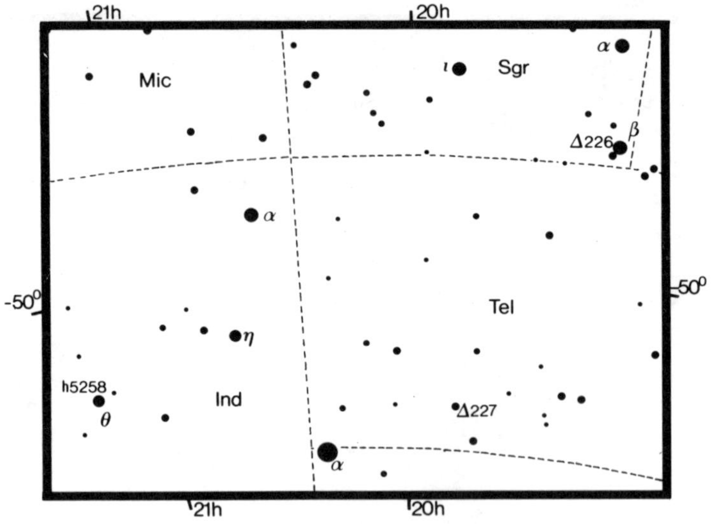

159 Object finder chart 23

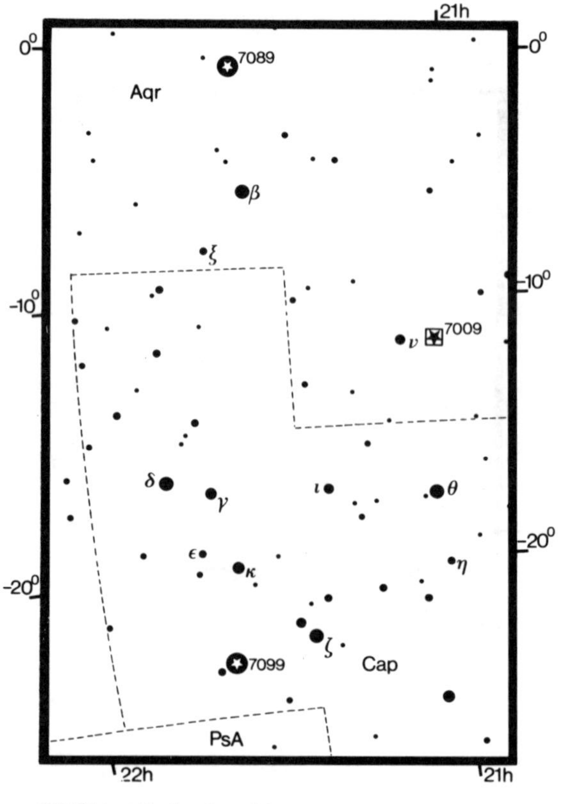

160 Object finder chart 24

Appendixes

I A SIMPLE METHOD OF CALCULATING THE JULIAN DATE FOR ANY DAY AFTER 15 OCTOBER, 1582

The function INT that appears in the formulae is the integer function. To use it you simply disregard the part of the number that comes after the decimal point. For example INT (2.4) = 2.

Firstly, you will need to know the Universal Time and date for your location.

Y = year, M = month, D = day, T = time.
If M = 1 or 2 calculate Y′ = Y − 1 and M′ = M + 12,
otherwise Y′ = Y and M′ = M.

Calculate A = INT(Y′/100)
Calculate B = 2 − A + INT (A/4)
Calculate C = INT(365.25 × Y′)
Calculate D = INT(30.6001 × (M′ + 1))

The Julian Date is then given by:
J.D. = B + C + D + (T/24) + 1 720 994.5

Example: Julian Date for 11:00am 4th January, 1987 at Sydney.

The Universal Time for this local time is 0 hours, 4/1/1987. You subtract 11 hours because Sydney was on daylight saving time. (See table 6.1)

Y = 1987, M = 1, D = 4, T = 0

Because M = 1 or 2 we calculate
Y′ = 1987 − 1 = 1986, M′ = 1 + 12 = 13.
A = INT(1987/100) = 19
B = 2 − 19 + INT(19/4) = −13
C = INT(365.25 × 1987) = 725 751
D = INT(30.6001 × (13 + 1)) = 428

J.D. = −13 + 725 751 + 428 + (0/24) + 1 720 994.5
= 2 447 160.5

A simple computer program can do this calculation for you. This program is written in BASIC and should run on most popular home computers, provided that they can handle numbers up to 3 000 000 as integers. Your computer's instruction manual will tell you if this is possible.

```
10 INPUT 'Enter Day';D
20 INPUT 'Enter Month';M
30 INPUT 'Enter Year';Y
40 INPUT 'Enter Universal Time';T
50 IF M < 3 THEN M = M + 12 : Y = Y - 1
60 A = INT(Y/100)
70 B = 2 - A + INT(A/4)
80C = INT(365.25 * Y)
90 D = INT(30.6001 * (M + 1))
100 JD = B + C + D + (T/24) + 1720994.5
110 PRINT 'The Julian Date is';JD
120 END
```

Calculating the Sidereal Time is a much more complex task. I have included the method here for those who feel it necessary to do their own calculations. A simpler way is to use one of the yearly publications which lists sidereal time such as The Astronomical Almanac or The Astronomical Society of New South Wales Ephemeris.

To start you will need to know the Julian Date at 0h U.T. and your longitude in decimal degrees.

These formulae expect times and angles to be decimal numbers, not the divisions of 60 that we usually use. To turn an angle into decimal degrees do the following:
Decimal degrees = degrees + minutes/60 + seconds/3600.

Calculate $T = (JD - 2\,415\,020)/36525$

The Sidereal Time at Greenwich for 0h U.T. is given by:
$ST = 24 \times FRC(0.276\,919\,398 + 100.002\,135\,9 \times T + 0.000\,001\,075 \times T^2)$

The function FRC is the fraction function. It means discard the part of the number before the decimal place. It is the opposite of the INT function.

To find the Sidereal Time at Greenwich for any time during the day add the following to the ST already calculated:
GST = ST + hours × 1.002 737 908 + minutes × 0.016 712 298 + seconds × 0.000 278 538

To find the local sidereal time LST you do the following to the GST just calculated:

LST = GST + (longitude/15) if your longitude is west.
LST = GST − (longitude/15) if your longitude is east.

The longitude must be in decimal degrees.

The LST given will be in decimal time—to convert this to normal time do the following:
The number before the decimal point is the hours.
Take the FRC of the time and multiply it by 60; the resulting part in front of the decimal point is the minutes. Take the FRC that is left now, again multiply this by 60; this gives you the seconds.

Example:

LST = 11.723 611
Hours = 11. FRC = 0.723 611
Multiply FRC by 60 to get 43.416 7
Minutes = 43. FRC = 0.416 7
Multiply FRC by 60 again to get 25
Seconds = 25.
So LST 11.723 611 equals 11h43m25s.

II USING THE SUN TO FIND NORTH AND SOUTH

The best way to find an accurate north-south line for astronomy is to use the sun. At noon, the sun is on the meridian, so any vertical object will cast a directly north-south shadow. So all you need is an exactly vertical object. I find the best thing to use is a plumb line as this will always hang vertically. So, simply hang a plumb line near where you want your N-S line and wait for noon.

Unfortunately, noon does not always occur at 12:00pm, for a number of reasons. The first of these is caused by the way we set our clocks. Because it is convenient to set all the clocks in an area to the same time, we choose a certain meridian for the standard. In Queensland, New South Wales, Victoria and Tasmania this standard is 150° east longitude. Unfortunately, very little of an area actually lies on its standard meridian, so the first correction we have to make is for this discrepancy. For every degree east of the standard meridian you must subtract 4 minutes from the time of noon, for every degree west, you must add 4 minutes. For example, Canberra has a longitude of 149° east; this is 1° west of the meridian, so you would have to add 4 minutes to the time of noon.

The second effect is astronomical. Because the Earth's orbit is not exactly circular, the Sun can run a little bit ahead of or behind its expected time. This difference can vary from sixteen and a half minutes fast in November to fourteen and a quarter minutes slow in February. The tables listing this discrepancy are found in The Astronomical Ephemeris and the ephemerides of local astronomical societies. For example, on 6 June 1987, the time of noon was one and a half minutes early. This combined with the four minutes we need to add for Canberra, gives us an actual time for noon of 12h02m30s.

Once you know the time of noon, wait until then, then mark two points on the shadow of the plumb line a few metres apart. Join these two points up with a straight edge and you have an exactly north-south line.

III CATEGORIES OF VARIABLE STARS

Regular Period Variables

Long period types

These have periods of between 120 and 1400 days. They vary in magnitude by up to six magnitudes.

Short period types

Eclipsing binaries. These are the close binaries; the variations are caused by the two stars eclipsing each other as they orbit. If you graph the magnitude of these stars a very interesting trace appears. The graph gives you the period of the close binary and some idea of the size of the two stars.

Pulsating. Pulsating stars actually change in size in a regular manner giving rise to the variations. They have periods of a few days and magnitude variations of one or two magnitudes.

Ellipsoidal. These are distorted stars that change in brightness by rotating and showing us a differing amount of surface. These stars are fairly rare.

Irregular Period Variables

Novae. A special case, these stars explode every so often throwing off the outer layers of their atmospheres. The increase in brightness of these stars is quite spectacular, rising by up to thirteen magnitudes.

Nova-like. These stars show sudden increases in brightness which are similar to novae, but do not involve the loss of surface layers. They increase in brightness by about three magnitudes.

Red giants. Some red giants will go through a number of pulsations in size as they change the types of fuel they are burning. This shows up as changes in brightness. Brightness changes are usually around one magnitude.

Miscellaneous

These are all the other stars that don't fit the main classes.

IV FORMULAE FOR VARIABLE STAR ESTIMATION

To use these formulae, you simply choose which of the three suits your observation, perform the calculation and record the result. The result is your estimation of the variable's magnitude. When you write your result, record only the first decimal place in the answer.

Star 1, the star which is brighter than the variable.
Star 2, the star which is fainter than the variable.

If the variable is half way between the two comparison stars, use this formula:

Variable magnitude = (Mag. Star 1 + Mag. Star 2)/2

If the variable is closer to the brighter star than the fainter, use this formula:

Variable magnitude = (3 × Mag. Star 1 + Mag. Star 2)/4

If the variable is closer to the fainter star than the brighter, use this formula:

Variable magnitude = (3 × Mag. Star 2 − Mag. Star 1)/4

Example:
If I estimated that a variable star lay between two comparison stars of magnitudes 8.9 and 9.4, and that it was close to the brighter star, the 8.9 one, I would do the following: Firstly I'd choose the right formula, in this case the second of those listed. I would then put the values in.

Mag. Star 1 = 8.9, Mag. Star 2 = 9.4,
Variable magnitude = (3 × 8.9 + 9.4)/4 = 9.0

I would record the variable as having a magnitude of 9.0 when I observed it.

V REPRESENTATIVE LIST OF ASTRONOMICAL SOCIETIES IN AUSTRALIA AND NEW ZEALAND

This list is not complete and we apologise to any societies we have omitted. Most societies welcome visitors to their meetings, though it is a good idea to contact the club in question beforehand, to make sure that arrangements haven't changed since this book was written. The brief descriptions which follow some of the societies' addresses were provided by the clubs themselves.

AUSTRALIA

Australian Capital Territory

Canberra Astronomical Society
Griffin Centre, Bunda St
Canberra City ACT 2601

The CAS, founded in 1969, caters to the needs and varied interests of the amateur astronomer. The society meets on the third Thursday of each month (except January) at the Jaegar Building, ANU, at 8.00pm. It also holds regular observing nights with the 225mm Oddie refractor telescope at Mt Stromlo. The society publishes a monthly journal, *The Southern Cross*, containing items of astronomical interest and members' observations.

New South Wales

Astronomical Society of New South Wales
PO Box 208
Eastwood NSW 2122

The ASNSW was founded in 1956 to foster an interest in astronomy amongst the public. At present the society has an active membership of over 300. In addition to holding the occasional viewing night for the public or schools, the society publishes a monthly magazine called *Universe* and a yearly ephemeris. The society is also nearing completion of its observatory on the outskirts of Sydney. Meetings are held on the second and fourth Wednesdays of each month at Burwood Public School, Railway Parade, Burwood at 7.30pm.

British Astronomical Association
Sydney Observatory
Sydney NSW 2000

The BAA accepts applications from all who would like to develop an interest in astronomy. Beginners are especially welcome. The Branch meets at 7.45pm on the third Wednesday of each month at Sydney Observatory. As well as a guest speaker who is a specialist in the field, the Branch holds regular viewing nights and practical workshops in astronomy.

Illawarra Astronomical Society
PO Box 1814
Wollongong NSW 2500

At the IAS we encourage new members through education and involvement in club activities. Our monthly observing nights, held under dark skies, are popular with our 'deep-sky' fans. Many members support observing programs such as grazing occultations and appulses. Meetings are lively and informal. They may include a speaker, news, reports, slides or a film. Meetings are held at the WEA Building, 97 Corrimal Street, Wollongong at 7.30pm on the second Saturday of each month. Contact the secretary at the above address. Visitors always welcome.

Sutherland Astronomical Society
PO Box 31
Sutherland NSW 2232

Wagga Wagga Astronomical Society
15 Castlereagh Ave
Wagga Wagga NSW 2650

Queensland

Astronomical Amateurs Club of Moreton Bay
116 Station Road
Indooroopilly Q4068

Astronomical Association of Queensland
PO Box 101
St Lucia Q 4067

The AAQ at present has more than 200 members. It meets monthly on a Saturday evening. Further details of membership and meetings can be obtained from the secretary at the above address.

Brisbane Astronomical Society
PO Box 204
Morningside Q 4170

The Brisbane Astronomical Society was formed recently to develop interest and share experiences with people who are fascinated by astronomy. It was founded by a group of amateur astronomers who have recognised the need for a society for the average person, where astronomy is explored at a practical level. The main emphasis is on 'hands-on' experience. The Society meets on the second Monday of each month at the Wynnum State High School Library. Meetings start at 7.30pm and finish around 10.00pm.

Bundaberg Astronomical Society
PO Box 586
Bundaberg Q 4670

The Society's observatory was opened in 1968 and contains a 49.5cm Newtonian telescope. The clubhouse contains a lecture room with an extensive book and photographic library. The club's computing section supplies all the mathematical data necessary for the observatory's operation. The society meets on the first Friday of each month in the clubhouse at 7.30pm; it also publishes a regular newsletter *The Alloway Observer*. Over the years the society has participated in programmes with the CSIRO, NASA and New Zealand societies.

Southern Astronomical Society
PO Box 685
Springwood Q 4127

The Southern Astronomical Society was formed by a group of amateur astronomers with an emphasis on observational astronomy. Many members have other interests ranging from photography to computing. Meetings are held monthly and members of the public are most welcome to attend. The society has a 320mm reflector telescope that is used by members and the public at field nights. For further information, please write to the secretary at the above address.

Sun Coast Astronomical Society
PO Box 25
Moffat Beach Q 4551

The society was formed in 1979 by an enthusiastic group of people sharing the common bond of an interest in astronomy. The society plans to establish an observatory on the Sunshine Coast for the benefit of members and the public. The Society meets at 7.30pm on the second Saturday of each month, in the science block of the Caloundra State High School, Queens Street, Caloundra. The society aims to pursue the research and knowledge of astronomy, to engage in its practical aspects, to cater for members and to promote the science of astronomy.

South Australia

Astronomical Society of South Australia
GPO Box 199
Adelaide SA 5001

Tasmania

Astronomical Society of Tasmania
c/- Norwood Avenue Post Office
Launceston Tas. 7250

The AST, founded in 1934, is a group of about 70 amateurs who share a keen interest in all facets of astronomy and related sciences. Meetings of the AST are held on the last Tuesday of each month from January to November. The usual venue is the South Hobart Primary School, Anglesea St, South Hobart, meetings commencing at 8.00pm and concluding at 10.00pm. Visitors are always welcome. Members receive a bi-monthly newsletter and the society produces a yearly ephemeris for Tasmania.

Victoria

Astronomical Society of Albury–Wodonga
1 Poplar Street
Wodonga Vic. 3690

The society was formed in February of 1974 and currently has a membership of 15 and a junior section of 8. The club's main work is the holding of public viewing nights and astronomy sessions for local schools and other interested groups. The society publishes a magazine, *Geodessey*, three or four times a year. Meetings, which often take the form of a viewing night or lecture by a guest speaker, are held on the third Wednesday of each month, though this is sometimes altered for special events.

Astronomical Society of Frankston
PO Box 596
Frankston Vic. 3199

Astronomical Society of Geelong
109a Ryrie Street
Geelong Vic. 3220

The Astronomical Society of Geelong was founded in September 1960. It operates from the Belmont Common in South Barwon. The current membership is about 35. For further information contact the secretary at the above address.

Astronomical Society of Victoria
GPO Box 1059J
Melbourne Vic. 3001

Founded in 1922, the ASV now has a membership of over 500 amateur astronomers in Victoria, around the country and overseas. The society meets at 8.00pm on the third Thursday of the month from February to November at the State Library and Museum. Access is from the east of the State Library in Latrobe Street, between Russell and Swanston Streets. Admission is free and visitors are most welcome.

Ballarat Astronomical Society
PO Box 284
Ballarat Vic. 3350

The Ballarat Astronomical Society was founded in 1959, and now has an active membership of about 40. The group's interests are varied and include radio astronomy, astrophotography and computing. The society has instruments ranging from a 125mm refractor to a 660mm reflector. The society meets at 8.00pm on the second Friday of each month at the Ballarat Municipal Observatory, corner of Magpie and Cobden Streets, Mt Pleasant, about 500m south of Sovereign Hill.

Latrobe Valley Astronomical Society
PO Box 144
Churchill Vic. 3842

The Latrobe Valley Astronomical Society is a small group of some 25 members who meet at 7.30pm on the second Monday of each month in Room IM207 of the Gippsland Institute of Advanced Education at Churchill. The Society has a 250mm reflecting telescope in its observatory at Yallourn. Members also meet here on selected nights each month. Interested school and other community groups are sometimes given a tour of the night sky by our members at the observatory.

Western Australia

Albany and District Astronomical Group
Lot 46 Bushby Road
Lower King via Albany WA 6330

Astronomical Society of the South West
23 Dunstan Street
Bunbury WA 6230

Astronomical Society of Western Australia
GPO Box S1460
Perth WA 6001

The Society, which was formed in 1949, has about 120 members. It meets at 8.00pm on the second Monday of each month in the Lesser Hall, Floreat Park Civic Centre, Howtree Place, Perth. The society has active astrophotography, deep-sky and meteor sections. Major activities include astrocamps and public field nights. A member has recently founded the *Australian Journal of Astronomy* to serve amateurs in the South Pacific Region.

NEW ZEALAND

North Island

Auckland Astronomical Society
Box 24-187 Royal Oaks
Auckland 4

Gisborne Astronomical Society
Box 678
Gisborne

Hamilton Astronomical Society
Box 9153
Hamilton

Hawera Astronomical Society
Box 190
Hawera

Hawkes Bay Astronomical Society
c/- Mrs J. Demanser
11 Bill Hercock Street
Napier

New Plymouth Astronomical Society
Box 818
New Plymouth

Rotorua Astronomical Society
Box 1621
Rotorua

Wairapa Astronomical Society
c/- B. Harvey
186D Columbo Road
Masterton

Wanganui Astronomical Society
c/- Mr B.S. O'Connell
17 Rata Street
Wanganui

Wellington Astronomical Society
Box 3085
Wellington

Wellington College Observatory
Wellington College
Mt Cook
Wellington

Whakatane Astronomical Society
21 Goulstone Road
Whakatane

Whangarei Astronomical Society
Box 436
Whangarei

South Island

Ashburton Branch
Canterbury Astronomical Society
20 Russell Avenue
Ashburton

Astronomical Section
Royal Society of New Zealand
Box 233
Nelson

Canterbury Astronomical Society
PO Box 25-137
Victoria Street
Christchurch

The Canterbury Astronomical Society has a membership of 150. Photometry and astrophotography are our main areas of interest although members are free to work in any area they wish. Overnight accommodation is available at our observing sites. The society has a 350mm Cassegrain and two 200mm telescopes.

Dunedin Astronomical Society
Box 6019
Dunedin

North Otago Astronomical Society
c/- W. Bradley
62 Tees Street
Oamaru

South Canterbury Astronomical Society
79 Arthur Street
Timaru

Southland Astronomical Society
Southland Museum
Box 1012
Invercargill

Glossary

Absolute magnitude The magnitude that a star, or any other object, would appear to be if it was 10 parsecs from the Earth

Achromatic This describes a lens which doesn't give a colour cast to its images

Albedo The measure of the amount of light an object reflects. For example, if a planet were to reflect 16% of the light falling on it, we would say that it had an albedo of 0.16.

Altazimuth mount A telescope mount that allows you to turn the telescope in two directions, from side to side, azimuth, and up and down, altitude. An altazimuth mount is the easiest to set up, but to follow a star through the sky you must constantly adjust the mount's position in both axes.

Altitude For astronomy this means the measurement in degrees of the position of an object above the horizon. It is part of one system for specifying the position of an object in the sky.

Aperture The diameter of the lens or mirror of a telescope

Aphelion The greatest distance of an object from the Sun. It is the opposite of perihelion.

Apparent magnitude The brightness of a star or object, as it appears in the sky. cf. absolute magnitude.

Asteroid One of the small minor planets that orbits between Mars and Jupiter. Asteroids are the remains of material left over from the formation of the solar system.

Astronomical Unit The average distance from the Earth to the Sun. It is approximately 149 600 000 km.

Averted vision A technique used by astronomers to see faint objects. It involves looking to one side of a faint object while concentrating on the object itself.

Azimuth This is the measurement of the direction an object appears to be in. To calculate it you draw an imaginary line from the object directly to the horizon below it; you then measure the compass direction of this calculated point, starting at north and working towards east. This angular measurement is the azimuth of the object.

Billion In this book we use the English form of billions, that is a one followed by twelve zeros. Some American books use billion to represent a one with nine zeros; in this book that is called a thousand million.

Cassegrain telescope A form of reflecting telescope that uses a spherical main mirror at the bottom of the tube and a small hyperboloidal secondary mirror at the top. The secondary mirror reflects the light back down the tube and through a small hole cut in the main mirror to the eyepiece.

Catadioptic This describes a telescope which uses both refraction and reflection to form an image. The popular Schmidt-Cassegrain telescopes are an example.

Celestial equator The line drawn in the sky which is equidistant from both celestial poles.

Celestial poles The points in the sky directly above the North and South poles of the Earth

Celestial sphere The apparent sphere of sky that surrounds the Earth. It relates to the ancient concept of the Earth being surrounded by a sphere containing the stars and planets.

Chromatic aberration An effect that lenses have on light passing through them. As the light is refracted it splits into its component colours as it does in a prism; this appears as coloured haloes around objects.

Cluster A group of stars. There are two main types, open clusters and globular clusters.

Collimation Having all the elements of a telescope aligned along the same optical axis.

Colour index A method astronomers use to record the colours of stars

Coma The region of gas surrounding the nucleus of a comet. It also describes a defect in a telescope where the image of a star appears to be a fuzzy blob with a small tail.

Comet A comet is a large mass of ice and dust which orbits the Sun. Comets are thought to be left over material from the formation of the solar system. When a comet approaches the Sun it forms a tail which may stretch over 100 million kilometres.

Constellation An area of sky defined by an agreement of the International Astronomical Union. All objects in the area are named after that constellation.

Dark adaptation When you first enter a dark area from a light area it takes some time for your eyes to adjust. The time taken is up to twenty minutes for complete dark adaptation.

Declination This is a measure of the angular distance an object is from the celestial equator. It is similar to the lines of latitude drawn around the Earth. Declination is one part of the coordinate system that is used by astronomers to specify the location of any object in the sky; right ascension is the other part—together they make up the equatorial coordinate system.

Deep-sky A term used to describe nebulae, clusters and galaxies. Objects which are further away than most stars.

Dichotomy The time when an astronomical object is exactly half illuminated. The only objects in the solar system that can show this to Earth are our Moon at first and last quarter and the planets Mercury and Venus when they are at greatest elongation.

Drive corrector A device which is attached to motorised telescope mounts to correct for inaccuracies in the mount or its alignment. Drive correctors are usually only needed for photographic work.

Eccentricity A measure of how squashed an ellipse appears.

Eclipse An eclipse appears when one object passes into the shadow of another. The two types of eclipses visible from Earth are lunar and solar eclipses.

Ecliptic The path followed by the Sun through the sky. All the planets and the Moon also appear to be near the ecliptic. The ecliptic passes through the twelve constellations of the zodiac and Ophiucus.

Elevation Elevation is the same as altitude

Ellipse An oval

Elongation The angular distance of one object from another. In astronomy the elongation normally refers to the distance an object appears to be from the Sun. For example, if a comet was 46° away from the Sun we would say that its elongation was 46°.

Epoch A date chosen to be the reference point for observations. The most commonly used one is Epoch 1950.0; it relates observations to the positions of the stars on 1 January 1950.

Equatorial mount A mount which allows your telescope to follow a star through the sky by turning it along only one axis. Equatorial mounts are necessary for you to do any photography with your telescope.

Equatorial coordinates A system of coordinates made of right ascension and declination. The coordinates rotate with the stars.

Filter Any of a variety of devices which are attached to eyepieces or telescope tubes to modify the light passing through them

Finder scope A small refracting telescope attached to a larger telescope to help in pointing the large telescope

First point of Aries The starting point of the right ascension system of coordinates. It is the point at which the Sun crosses the celestial equator in March each year.

Focal length The distance from a lens or mirror to its focus

Focal ratio The focal length of a lens or mirror divided by its aperture. In photography, this refers to the speed of a camera lens.

Focus The point where all the converging light rays from a lens or mirror meet

Galaxy A large disk or ball of millions of stars and nebulae. Galaxies are the largest individual structures in the universe.

Gegenschein A dull glow opposite the sun in the sky caused by small particles of dust scattered through the solar system.

Globular cluster A huge sphere containing thousands of stars. Globular clusters are observed to surround our galaxy and many other nearby spiral galaxies.

HII region HII, pronounced 'H two' is the ionised state of hydrogen gas. It exists in emission nebulae as the glowing gas we see. A HII region is simply another way of saying emission nebula.

Hubble's Law This is a theory developed by Edwin Hubble, and refined by others that relates the distance of an object to its speed of recession. It is one of the basic parts of the expanding universe theory.

Hypersensitisation A chemical means of increasing the suitability of photographic film for astronomical work

Inclination The angle that the orbit of one astronomical body makes with the orbit of another body

Julian Date A system of counting days used by astronomers to record the date of an observation. It corresponds to the number of days since 1 January 4713 B.C.

Lens A specially shaped piece of glass or other transparent material that will bend light so that parallel beams come together at the same point, the focus. Lenses come in all shapes and sizes, the most common forms being shown in figure 32.

Light-year The distance that a beam of light travels in one year. It is approximately 9 460 529 700 000 km.

Magnitude This is the system astronomers use to measure the brightness of stars. It was originally based on the Babylonian system of having six main classes of stars, but it has been refined by scientists since. The basis of the system is that each magnitude is a certain number of times brighter than the magnitude above it. The exact scale for the magnitude system is the fifth root of one hundred, or about 2.51.

Mare, Maria The large dark patches on the Moon. Mare (pron. mar-ay) is the Latin word for 'sea'.

Meridian The local meridian is the line running directly overhead from north to south.

Meteor A small particle of dust which enters the Earth's atmosphere at high speed and burns up because of the high temperatures generated. Meteors are thought to be particles left over from the passage of comets.

Meteorite A larger body than a meteor which survives its passage through the atmosphere to reach the Earth. Rather than comets, the asteroids are thought to be the source of meteorites.

Nadir The point directly beneath your feet, that is directly opposite the zenith

Nebula A large cloud of gas. Some nebulae are visible because of embedded or nearby bright stars, the emission and reflection nebulae. Other nebulae are dark and are seen only by the way they obscure the light of objects behind them.

Newtonian telescope A reflecting telescope based on an original design by Isaac Newton. It uses a paraboloidal primary mirror at the bottom of the tube and a small flat secondary mirror to direct the light through a hole in the side of the tube, where it can be viewed with the eyepiece. Newtonian telescopes are the most common form of small reflecting telescopes.

Nova A star which suddenly increases its brightness

Nucleus A nucleus is a large ball of ices and dust which forms the central part of a comet.

Object glass This is another name for the main lens of a refracting telescope.

Occultation An occultation happens when one astronomical body passes in front of another.

Oort's Cloud The cloud of comets which is thought to surround the solar system at a distance of two light-years

Orbit The path followed by one body as it moves around another

Perihelion The closest that an object gets to the Sun

Precession The motion of the stars and planets which causes them to slowly change their equatorial coordinates. Precession results from the Earth wobbling on its axis every 25 800 years.

Radiant The point from which meteor showers appear to originate. The radiant is purely an effect of perspective.

Reflecting telescope A telescope that uses a specially curved mirror to form an image of the objects it is pointed at. The first reflecting telescope was made by Sir Isaac Newton, the English scientist, in 1671.

Refracting telescope A telescope that uses a lens or system of lenses to form an image of the object it is pointed at. The refracting telescope was invented in 1609 by a Dutch spectacle maker, Hans Lippershey.

Reticle A system of lines and angles ruled in an eyepiece so that you can measure separations and angles between objects.

Right ascension This is a measure of the angular distance of an object from an imaginary line in the sky. Right ascension is similar to the lines of longitude drawn around the Earth. Together with declination, right ascension forms the equatorial coordinate system, the system astronomers use to specify the position of any object in the sky.

Schmidt telescopes These telescopes are similar to Cassegrain types. They have a correcting plate to make them suitable for photographic work. It is possible to buy Schmidt cameras that are specially designed for high quality photography.

Scientific notation This is a shorthand method of writing very large or very small numbers used by scientists. Instead of writing all the zeros in a number, a scientist will count them up and write them as a power of ten. For example, one billion can be written as 1 000 000 000 000 or as 1×10^{12} in scientific notation. A small number such as one trillionth can be written either as 0.000 000 000 000 000 001 or 1×10^{-18}.

Seeing A method used to describe the clarity and stillness of the night air when making observations

Sidereal time The method of time keeping which uses the motion of the stars rather than the Sun. Twenty-four hours of sidereal time is the same as 23h56m4s of normal solar time.

Solar wind The stream of sub-atomic particles that the Sun throws out into space. These particles cause the Aurorae Australis and Borealis when they reach the Earth.

Spectrum The spread of colours which makes up the light from distant stars. By studying the make-up of the spectrum, scientists can tell what materials were present when the light was formed.

Supernova A rare astronomical event in which a massive star explodes at the end of its life. Supernovae are so bright that they briefly outshine all the other stars in their galaxy before they fade from view.

Tail The gases and dust left behind by a comet in its orbit about the Sun. The tails of comets can stretch over 100 million kilometres.

Transit The movement of one astronomical body across another or through some point in space

Twilight The time between sunset or sunrise and when astronomers consider the sky dark enough to begin observing. Astronomical twilight is when the sun is less than 18° below the horizon. Nautical twilight is when the sun is less than 12° below; civil twilight, less than 6°

Zenith The point directly overhead, that is 90° altitude

Zodiac The constellations along the ecliptic

Zodiacal light A band of light which stretches along the ecliptic. It is caused by small particles of dust stretching along the Earth's orbit.

Bibliography

Astronomical Society of New South Wales *Ephemeris 1987* Eastwood: Astronomical Society of New South Wales Incorporated, 1986

Bartholomew, J.C. et. al. *The Times Atlas of the World* 7th edn, London: Times Books Limited, 1985

Becvar, A. *Atlas Australis* 2nd edn, ed. prof. dr. Mohr J.M., Cambridge: Sky Publishing Corporation, 1976

—— *Atlas Eclipticalis* 2nd edn, ed. prof. dr. Mohr J.M., Cambridge: Sky Publishing Corporation, 1974

—— *Atlas of the Heavens II* Catalogue 1950.0. 4th edn, Cambridge: Sky Publishing Corporation, 1964

Burnham, R. jnr *Burnham's Celestial Handbook* vols 1, 2 & 3, 2nd edn, New York: Dover Publications, 1978

Covington, M. *Astrophotography for the Amateur* Cambridge: Cambridge University Press, 1985

Duffett-Smith, P. *Practical Astronomy with your Calculator* 2nd edn, Cambridge: Cambridge University Press, 1981

Dunlop, J. 'A Catalogue of Nebulae and Clusters of Stars in the Southern Hemisphere Observed at Parramatta in New South Wales' *Memoirs of the Astronomical Society of London* vol. 3, 20 December 1827, pp. 113–151.

——'Approximate Places of Double Stars in the Southern Hemisphere Observed at Parramatta in New South Wales' *Memoirs of the Astronomical Society of London* vol. 3, 9 May 1828, pp. 257–275

Gore, R. 'The Planets' *National Geographic* vol. 167, no. 1, January 1985, pp. 4-51

—— 'What Voyager Saw: Jupiter's Dazzling Realm' *National Geographic* vol. 157, no. 1, January 1980, pp. 2–29

—— 'Uranus' *National Geographic* vol. 170, no. 2, August 1986, pp. 178–195

Hartung, E.J. *Astronomical Objects for Southern Telescopes* Melbourne: Melbourne University Press, 1984

Her Majesty's Nautical Almanac Office *The Astronomical Ephemeris for the year 1967* 200th edn, London: Her Majesty's Stationery Office, 1965

—— *The Astronomical Ephemeries for the year 1977* 210th edn, London: Her Majesty's Stationery Office, 1975

Herschel, Sir John F.W. *Results of Observations Made During the Years 1834, 5, 6, 7, 8* Cornhill: Smith, Elder & Co., 1847

Howse, D. *Greenwich time* Oxford: Oxford University Press, 1980

Jeffers, H.M., van de Bos, W.H. and Greeby, F.M. *Index Catalogue of Visual Double Stars* 1961.0 parts 1 & 2, Publications of the Lick Observatory vol. xxi, Mt Hamilton: University of California, 1963

Malin, D. and Murden, P. *Colours of Stars* Cambridge: Cambridge University Press, 1984

Mallas, J.H. and Kreimer, E. *The Messier Album* Cambridge: Sky Publishing Corporation, 1978

Murden, P. and Allen, D. *Catalogue of the Universe* Cambridge: Cambridge University Press, 1979

Meeus, J. *Astronomical Formulae for Calculators* 2nd edn, Richmond: William-Bell, 1982

Moore, P. *The Atlas of the Universe*, Sydney: Reader's Digest Services, 1970

—— *Guide to the Moon* Guildford and London: Lutterworth Press, 1976

Norton, A.P. *Norton's Star Atlas* 17th edn, ed. Satterthwaite, G.E., Edinburgh: Gall & Inglis, 1978

Sidgwick, J.B. *Amateur Astronomer's Handbook* 3rd edn, prep. Gamble, R.C., New York: Dover Publications, 1971

—— *Amateur Astronomer's Handbook* 4th edn, prep. Muirden, J., London: Pelham Books, 1979

—— *Observational Astronomy for Amateurs* 3rd edn, prep. Gamble, R.C., New York: Dover Publications, 1971

United States Naval Observatory & Royal Greenwich Observatory *The Astronomical Almanac 1986* Washington: US Government Printing Office, 1985

Wilson, C. *Starseekers* London: Hodder & Stoughton Ltd, 1980

Index

Entries which refer to figures or tables are in *italic* type.